THE BOOK OF THE HACKLE

THE BOOK OF THE HACKLE

FRANK ELDER

1979

SCOTTISH ACADEMIC PRESS
EDINBURGH

Published by
Scottish Academic Press
33 Montgomery Street
Edinburgh EH7 5JX

SBN 7073 0223 4

Printed in Great Britain by
R. & R. Clark Ltd, Edinburgh

Introduction

My father's interest in dry-fly fishing for trout developed within him an interest in breeding poultry to provide the feathers which have the essential qualities necessary to keep the artificial fly floating on the surface of the water. In the last fifteen years or so of his life, during which he bred poultry exclusively for this purpose, he was forced to ask fundamental questions not only about how different characteristics and colours could be produced through selective breeding but more fundamentally what exactly is a "good" hackle and what purpose is the hackle on a dry-fly intended to fulfil. The results of these enquiries derived from observation and practical experience are contained in this book.

At the end of Chapter I my father writes that it was unfortunate for anglers that Dr William Baigent did not live long enough to write his book. It is fortunate indeed that when my father died in December 1977, the book upon which he had devoted so much time and effort over the years was already with the publishers in proof form, complete with the magnificent paintings by Ellen Chapman of the feathers supplied by my father.

I am indebted to those people and in particular to Jack Heddon and Timothy Benn who have helped me to bring *The Book of the Hackle* the last short distance to publication.

IAN ELDER

Contents

List of Illustrations

Foreword

Judging by the articles and correspondence in the fishing magazines of today compared with those published round about the turn of the century, it is obvious that in these days there was a much greater interest in hackles, their quality and colour, than there is today. Articles dealing with some aspect of hackles appeared at frequent intervals and invariably correspondence on the subject followed in subsequent editions. Today, there are many descriptions of new fly-dressings, but the hackle, the essential ingredient in every dry-fly, is ignored. If an article on hackles does appear, it passes without comment.

The reason for this change is not clear. Possibly it is because in these days, dry-fly fishing, as we know it, was new. Fishing dry was the 'done' thing and every aspect was discussed and considered. I suspect, however, that, as I have attempted to show in this book, contrary to what is normally accepted, it is now much easier to obtain a good supply of dry-fly hackles. Today anyone can purchase hackles of good average quality from one of the many firms which stock fly-tying materials. With good hackles easily available, there is no need to search and so the interest has faded. Admittedly we do not have the range of colour available in earlier days. Today I don't suppose one fly-dresser in a hundred has ever seen a natural blue dun and although our average quality is good, I doubt whether the normal fly-dresser of today can appreciate the difference between this and a really fine hackle. The result is that for the majority there is now no search for the rare colours or the finest quality. The interest and the quest have ceased.

There is, I am glad to say, still a number who look for something better than the 'run of the mill' and I hope that this book will show fly-tyers, particularly those who have recently started to dress their own flies, that while the commercial hackle is good, something better does exist and that it is worth searching for. If, as Mr Sam Harris says, the commercial supply of good capes slowly dries up, the standard will deteriorate, the good will become moderate and then bad. If this happens, then I would expect the interest in hackles to reawaken. To tie a fly takes time and the real reward is to produce something which pleases the eye. A dry-fly tied with a bad hackle will always be a bad fly.

Although the hackle is the essential part of a dry-fly, it is amazing that it has never been examined in depth. Thomas Hughes wrote about breeding poultry for hackles. Dr Baigent did not live to write the book which was to be published with his collection of hackles. John Henderson came the nearest to it, for his articles, published mainly in the *Journal of the Flyfishers' Club*, cover a number of aspects and he also bred poultry to produce hackles. In this country these are the only three.

Practically every book on fly-fishing says something about hackles, but when I started to study the subject I soon realised that much of what was written was simply repeating what had been said before and much of it would not stand up to close examination. In this book, I have tried to put down as fact only the things that I have tested and found to be true.

To understand hackles, it is essential to have bred poultry for them. This is difficult to explain, but if you look only at capes and hackles, you are dealing with a dead, not a live, subject and you are only dealing with a finished product. To know all about roses you must grow them and it is the same with hackles.

As I became involved in my breeding programme, it was gratifying how many people came to my aid. Those with detailed knowledge of a certain subject went to the most extraordinary lengths to provide me with the information that I required and poultry experts did everything possible to produce the stock that I needed. If I tried to name everyone who had helped me, the list would be enormous, but I do acknowledge my sincere thanks to Mrs Robert Elder of East Bearford, who looked after my first birds and has kept a stock going ever since, and to Wilfred Crawford, who allowed me to carry on my breeding programme at Ugston for many years. Lacking the ground and facilities myself, I could never have built up my strain without their help.

In the early days collecting stock with the correct qualities was not easy and without the help of William Langton of Kingswells and James McEwan of Lumsden, well known poultry breeders, it would have been impossible. Jack Heddon's assistance was invaluable in covering the back literature. Nor would this book have been possible without the help of the East Lothian County Library, who searched the whole of the British Isles to produce the books that I required.

John Henderson and Timothy Benn, both of whom have bred game cocks for hackles, examined the results of each year's breeding and their suggestions have been most helpful.

I hope that this book will help fly-tyers to a better understanding of hackles and to appreciate the difference between the good, bad and indifferent.

North Berwick
1977

CHAPTER I

How it began

My special interest in hackles, which in turn induced me to breed poultry in an attempt to produce fine dry-fly hackles, came about fortuitously. The original idea was simply to produce enough of the rare colours for my own use, but this in turn quickly led to an intensive study of the subject and the building up of quite a large breeding stock. Although I have fished all my life, before the Second World War my hobby was ornithology. I have always been interested in birds and almost all my spare time was taken up with bird-watching. During the war I spent one or two leaves from the army fishing in Scotland with an old friend and at the end of the war, when I was stationed in north Germany waiting to get back to civilian life, I started to tie flies. Taverner's book *Fly-Tying for Trout*[1] was recommended and I could not have started with a better book. The Royal Engineers produced a fly-tying vice in their workshop, the only vice that I have ever owned and still use. There was very good shooting round Oldenberg. A wonderful marshy area contained almost every European variety of duck; Storks bred on the surrounding roof tops, and I have watched parties of Ruffs displaying in the Spring. In the Autumn we shot many duck and other game, so my fly-tying kit grew quickly. Every bird that came in was looked over by me in search of feathers, many of which I have to this day. I can produce the wings of a Jack Snipe or Garganey, though I cannot believe that I shall ever use them.

After the war, my interest in fishing increased and in 1960 I started to fish for a week every Spring with Dr Derrick Morton, a dedicated fisherman, and it is due to his influence that my interest in fishing, and dry-fly fishing in particular, has increased. In 1961 we went to the Aberdeenshire Don, and from that date I have fished on that delightful river every year. The Don is not a salmon river of the quality of its famous sister the Dee; it is a smaller river and the Don valley is wider and richer agriculturally. In farming there is an old saying, 'A mile of Don is worth two of Dee.' The reason is that the Don valley is not the acid peat of the Dee; old lime kilns lie up and down the strath, showing that lime is near the surface, and I am sure that it is because of this that the Don is such a fine trout river! The runs and pools lend themselves to enjoyable dry-fly fishing, and the runs are not too fast, which demands incessant casting, yet the pools have enough run to float the fly over the trout.

In the Forbes Arms at Bridge of Alford the picture of Dr William Baigent of Northallerton photographed at his fly-tying desk hangs above the tremendous sideboard at the end of the dining room. I soon found that 'Baigent's Brown' was *the* fly on the Don. It was automatically presumed that you fished it, so not knowing the correct tie, I purchased a supply for the following year. I was sent a yellow-bodied fly with a very, very long, sparsely tied, furnace hackle and I could do nothing with it. There was no doubt that a new fly cast over a rising fish usually had an immediate response, but after a few casts the fly looked like a drowned spider, any amount of 'fly floatant' failed to help and I quickly found that the Don trout, when a hatch was on, refused to touch any fly that was not riding right on the top of the water. I was told that Dr Baigent had kept Old English Game cocks to provide himself with hackles, so I decided that this must be the secret. A really stiff game cock hackle was the answer.

When I look back on this, I now realise how little I knew about hackles in these days. I had been fly-tying for a good number of years and had read almost all the books on the subject, but there is no book which describes the essential requirements of a really good hackle. Had I known anything about the subject, I would have realised that no hackle ever produced by any cock could support a hook when tied in this manner. I hasten to say that I now know that it was not the Doctor who was wrong; the flies normally sold as Baigent Browns are not as he tied them.

My lack of knowledge, however, had a good result. I decided that I would have to try to obtain hackles of quality, so that I could produce flies of the Baigent pattern. Really fine hackles cannot be obtained, so I thought that I would start to breed some for myself. The books, including Taverner's, all say that the Andalusian produces the fine steely blue colour, which all fly-tyers are looking for, but it seemed to be generally agreed that the Old English Game cock, the fighting cock used in the cock pit, produced the finest quality of hackles and this breed, unlike the Andalusian, has the added advantage that it can be obtained in every colour known in poultry. The fighting cock was the bird, but before I started a flock, I thought that the sensible thing to do was to read what those who had kept poultry for hackles in the past had to tell us. Keeping cocks for hackles was not uncommon round the turn of the century and I thought that articles by some of these breeders would tell me how to produce the colours I required. To my surprise I found that there were only two people who had written anything of importance on the subject. Thomas Hughes who lived in Wales and whose articles published in the *Fishing Gazette* in 1922[2] are the most authoritative. He kept Old English Game for some forty years. Later he wrote other articles on the subject, which are published in *Salmon and Trout*[3] and *Feathered World*,[4] but his articles of 1922 are the fullest and the best. The other man who, when he lived in Wales, bred cocks for this purpose is John Henderson and he wrote a number of articles on the subject in the *Flyfishers' Journal*[5,6] and the *Fishing Gazette*.[7] He is now an old man living in London, but is still extremely interested in the whole subject of hackles. I cannot exaggerate the importance of the advice and suggestions that he has given to me. He remembers many of the great anglers before and after the First World War and I have a most treasured collection of hackles which he has sent me; they come from breeders all over the world. Each Winter I have sent him a range of samples taken from that year's breeding and his critical appraisal is of tremendous importance to me. Without the judgement of an outside expert it is very difficult to evaluate your own work and it is easy to continue along the wrong line in selection and breeding. I am fortunate too in having Timothy Benn, who, having also bred cocks for hackles, gives me his opinion on each year's results.

If anyone starts to breed poultry for hackles, I would advise them to read the articles written by Hughes and Henderson. Although I have tried to include the important points that they make in my chapter Breeding for Hackles, it is certainly worth reading the originals.

I obtained a clutch of true Old English Game with some difficulty. It is now a rare bird, but at the present time there are at least two societies, the Oxford and the Carlisle, which look after the breed and hold annual shows. In addition, the sport of cock-fighting is not dead in certain parts of this country. I realised that I would have to breed and select to obtain the colours that I wanted, but I imagined from what I had read in the angling books, that the quality and shape of the hackle from this breed would be perfect. I quickly found my error. An Old English Game can produce a hackle of high quality, but it can also produce a very poor one. As I have shown in my chapter on breeding, you have to breed for quality, shape and colour, which makes the whole business very, very complicated.

Once I had started to produce hackles and was able to appreciate the real differences between a good and a bad one, I began to realise that out of the thousands of books which have been written on fly-fishing, with one or two exceptions, nothing of any real importance has ever been written on hackles. The hackle is the most important part of any trout fly and every book and article dealing with flies usually has a few words or sentences on hackles. A typical example written in 1946 is: 'By far the most difficult problem confronting the fly dresser is to obtain first class cock hackles. These are the neck feathers of the bird and should be long in the stem, short in the fibre, glossy, the same colour both sides and have a good natural springiness.' One could find almost the same thing written in hundreds of books. When you study it, you realise that it is saying nothing that is of practical use. It is all reasonably true, but unless you have handled many hackles and have sufficient experience, how do you know whether a hackle is long in the stem and short in the fibre? You have no yardstick against which you can measure the quality. It is a matter of standards and how critical you are in your assessment. A hackle which one fly-dresser would call very good would be discarded by another. When I started to study the subject in depth, I began to realise that there ought to be some standard against which one could gauge a hackle. In angling books written a hundred years ago you will read that good hackles are now almost impossible to obtain and that in the past they were common. Is it true? How fine were these magnificent hackles which came from the fighting cocks in the days before cock-fighting was made illegal?

Strange to say, in all the books written about every facet of angling, there has never been a book written on the hackle, this all-important part of any dry-fly. Many books skate over the subject without dealing with it in any detail and the same thing is repeated in book after book. The only person who started to write a book on the subject was Dr Baigent and unfortunately he died before he had time to write the text, but at least his collection of beautiful hackles which were intended to illustrate the book were privately published. Because it was his influence which made this book possible and because I consider that a detailed consideration of his study on hackles is important, I have included a fuller account of his work.

When I realised the need for a book which dealt with all aspects of the hackle, I started to collect material with a view to filling this gap. It was immediately apparent that if this book was to be written, it must cover the subject in depth and this meant that angling literature from the beginning would have to be covered, to find out what the authors of the past had to say on this subject. The fact that there has never been a book on hackles made the task even more formidable, for if there had been such a book it would no doubt have covered the literature to that date and it would thus have pin-pointed the important references. Without such a book it means a search of the literature from 'The Boke of St Albans' onwards. I have not the knowledge or the time to undertake such a mammoth task, but as in all my work with game cocks, I have been more than fortunate in that my interest in hackles has brought me in touch with someone who was able and willing to help. In this case it was J. S. Heddon. A study of angling literature is Jack Heddon's hobby and there can be few if any in Britain with his knowledge of the subject. When we discussed my proposed book, he volunteered to search the literature and to send me the references to any mention of hackles. This information allowed me to cover the essential material without having to search. At first he concentrated on books such as Halford, Cutcliffe, Skues, where he knew that he would find important facts on hackles, but after the normal books had been covered, his references came from more and more obscure publications. How he managed to cover so many books I do not know. He realises that amongst the immense number of angling books that have been published, it is possible that he has missed a book or article which has something really interesting to say on the hackle, but I feel confident that he has supplied all the references of importance. Without his help this book could not have been written.

I have mentioned that every book for the past hundred years has stated that the quality of hackles is not as good as it used to be, but is it correct to say that the hackles in the days long ago, when fighting cocks were common, were indeed far better than the hackles of today? While I am sure that the average quality may have been better, I do not think that the best of these days was better than the best of today. In a choice of colours, I am sure that they were much more fortunate. Blues, which are now almost unknown, have always been rare but all the other colours would be easily obtained. In quality and shape I do not think that they had anything better than our best. Without a standard, it would be impossible to judge, but fortunately there is a yardstick. There is the magnificent collection of fly-tying materials belonging to the Flyfishers' Club,[8] which allows us to see what was available in the past. This amazing collection took four years to complete and the Club's Annual Report of 1899 confirms that the sub-committee who brought this collection together had searched the whole of Britain to obtain the best samples. Many of the specimen hackles would come from fly-dressers' collections. I am sure that some of them would not be new, they would be hackles taken from cocks many years before and prized as perfect examples. They would be given to make this club collection as complete as possible. I think we can safely say that this collection truly represents the finest obtainable in colour and quality since at least early in the nineteenth century. In addition to the full collection of cock's hackles, there are sections to show representative samples from Wales and Devon, for these two areas were renowned for their quality of hackles. Indeed if you read the *Fishing Gazette* of the late 1800s you will find references to the quality of the Devonshire hackles. There is even a series of cocks' hackles taken from Dr Cutcliffe's dubbing book. In this collection we have actual examples of the finest that has ever been produced in hackles.

Anglers are fortunate that the suggestion put forward by the President at the Annual Meeting of the Flyfishers' Club in 1895 was accepted. The proposal by Mr Marston was that the Club should compile a collection of fly-tying materials 'for the interest of present and future members'. The sub-committee which was formed carried out an amazing task. The hackles, both cock and hen, form but a small part of the collection. The aim was to show a sample of every material, fur or feather, used in fly-tying, and the committee reported to the Club in 1899 that they had completed their task, the only items which they had been unable to obtain being the scapulars of a Roseate Tern and the fur of a Marten. I am glad to say that after all these years Jack Heddon has now been able to give the Club a sample of Marten's fur. Now that the Roseate Tern is once more a breeding species in Britain, perhaps someone finding a dead bird will complete the work of the committee. I have a feeling that Skues, who was on that sub-committee, was the driving force and we are grateful that they carried out such a wonderful task. Without this collection, the quality of hackles in the nineteenth century would be a matter for conjecture and no one would know whether there was or was not truth in the saying, 'In the old days hackles were far better than anything which can now be obtained.' Having studied this collection, I am of the opinion that while they had a magnificent range of colours, the quality and shape was no better than the best at the present time; indeed, I have hackles of my own breeding, which are at least as good for quality and in my view of better shape.

William Baigent, M.D.

William Baigent was born in 1864. He must have been a passionate fisherman from his earliest years, for his daughter, Mrs Lodge, has a notebook containing an essay on 'The Trout' written in his own hand in 1878. The account of the various flies of interest to the trout is illustrated by hand-coloured drawings of the bodies and legs and the actual wing from the fly has been pasted on top in the correct position. Alongside many there is an artificial which he tied to represent the natural.

It is clear therefore that from early in his life he was studying the natural flies that were the food of the trout and to my mind his great strength was that he did not copy anyone, mainly because there was no one to copy. No doubt he fished a wet-fly in his youth, but he quickly changed to a dry-fly and for the great part of his angling life he was a dry-fly fisherman. Although the art of fishing a fly on the top of the water is probably as old as the hills, the dry-fly as we now know it, always fished upstream, was made popular by Halford and it was this method that appealed to the young William Baigent. We gather from his letters to the *Fishing Gazette* that he was on his own in the north of England to use this method about the year 1890; and there was no one with whom he could discuss flies and fly-dressings. It followed that he produced his own, based on representation of the natural insect, but also, and here surely is the great value of his dressings, on what, by practical experience, he found would interest the trout on the waters he fished. Halford produced exact imitations of the natural fly. He had the fly in front of him and with, to use Skues' words, 'silk, fur and feather', he produced a representation which was as accurate as his fingers and materials could create. But did it interest the trout? That is the essence of the matter and I think that there are very few of Halford's ties which are religiously followed today. Baigent, to me, tackled the subject in a practical manner. He knew his insects. Even his youthful notebook tells us that, but his series of flies was not a slavish imitation of the insect as the human eye sees it lying dead on the fly-dressing table. His flies are the result of many years of trial and error, starting almost certainly from an exact representation as seen by the human eye and evolving through practical experience to a fly which would catch trout. Indeed we know that this is correct, for at one time when asked what his Baigent Brown was tied to represent he replied, 'It is not tied to represent any fly, it is tied to catch trout.'

The whole concept of the Baigent Brown and the other flies in his series clearly shows the careful study of the fly and the trout on the rivers and under the conditions that he fished. I carefully add this, for it is possible, indeed probable, that under different conditions of river, light, temperature or season, the flies which were so successful for him might be of little use in a different environment. I am sure that he would have agreed. His flies were evolved to produce an answer to a particular problem. Given a different problem, his answer would have been different.

Had Dr Baigent lived long enough to write the book which he had planned to accompany his series of hackles, his name would be far better known than it is today and his place in angling history would be acknowledged. His 'book' was published in 1937 by his widow as a limited edition of 70/80 copies. It is a scarce volume and as it has now become a collector's book, the normal angler or fly-dresser will have little chance of seeing a copy. It is therefore worth describing, for up to the publication of my book, it is the only one dealing with hackles alone that has ever been attempted. The 'book' is a stiff-backed folder or box made to look like a book. When opened up there are eleven loose, thick, cardboard sheets which are cut out in the middle like the cardboard mount surrounding a picture. Each of these mounts is actually made of two sheets of card glued together and in the cut-out opening is a row of actual hackles, their stems fixed between the glued cards. By mounting them in this way, the back as well as the front of the hackle can be examined, the strength of the feather can be tested with the fingers and by holding the card at an angle against a dark background the full colour and sheen of the hackle can be seen. In each book there are a total of 164 hackles, each one numbered and the colours named. Numbers 1–124 are natural cock hackles, the total being made up with dyed, hens', young cocks', saddle and spade hackles.

I have been fortunate in being able to examine not only the hackles in the book, but many of the capes from which the hackles were taken, for these are still in the possession of Mrs Lodge. The colours of the capes are superb. Although Dr Baigent kept Old English Game for many years, he left no notes on his breeding methods and there is nothing to help anyone attempting the same thing. According to Mrs Lodge, he used Andalusian hens frequently, no doubt to obtain the blue colour. Having a country practice, he was able to scatter his birds over the countryside, to be looked after by the local farmers, and in these days when every farm had its barnyard fowls, they would be welcome to improve the eating quality of the chickens bred from them. A very satisfactory arrangement for both parties. Over the years he must have had a very large number of birds judging from the range of colours he produced in his book, for I know from experience, that to be able to produce this range, he must have had a very, very large number of capes from which to choose.

The colours of the hackles are superb. In the rusty blues in particular, he has an amazing series of shades, all quite outstanding. He was particularly fond of the colour which he called 'blue red'. This is obtained by crossing blue and red. It is a red, but the shade of red tends towards the blue not the yellow side of the spectrum. In a dark red it could be called blood red. The reds are not on the whole popular colours for hackles at the present time and you rarely see a modern tie, particularly for a dry-fly, which specifies a red hackle, but Baigent considered the blue reds to be the most valuable of all hackles for catching trout. He said that if a blue red could not be obtained, a red hackle with a blue one tied through it was the best answer. Another doctor, Dr Cutcliffe, also put great faith in a red hackle, so perhaps this colour should receive more attention from the modern fly-dresser. The fact that this colour is usually the finest for quality is also a good reason for its use where possible.

The quality of the hackles bred by Dr Baigent is good, although perhaps not brilliant, possibly due to the use of the Andalusian, which often tends to be soft. The shape is definitely poor by modern standards but, as I have shown in my chapter on shape, Dr Baigent did not put great emphasis on shape and he would not select for this character.

In the 1920's he produced his 'Refracta' series of flies. The reason for this series is not clear and I do not think that they have any great value. The idea appears to have been that the fly should be tied with two hackles, one much longer than the other. The longer hackle supported the fly and made the correct pattern on the surface, while the trout looked at the shorter hackle, which was of a different colour and represented the wings. The fact that the trout's line of sight was refracted as it passes through the surface was supposed to be of some importance in the resulting image of the hackle to the trout. Exactly what was the principle behind it is not clear to me. This series never gained any popularity and is never seen today. Mr J. J. Hardy was a friend of Dr Baigent and I have a feeling that it was his commercial interest that persuaded Dr Baigent to produce this series. Messrs Hardy of Alnwick were given the sole right to market these flies.

In 1967 I wrote an article on 'Dr Baigent's Flies'[9] for the *Journal of the Flyfishers' Club* and Messrs Hardy kindly allowed me to give a full list of the dressings. Dr Baigent had given them to this firm and to my knowledge they had never before been published. When examining all the documents kept by Mrs Lodge, I found a complete list of the dressings written in the Doctor's own hand. In a number of ways these dressings differ from those given in my article. From the way that they were worded, it would appear that they were written with a view to publication, but I do not think that they ever appeared in print at that time.

When he wrote these notes is not known, for they are not dated, but the fly that I have examined, which he tied in his later years for use on the Don, certainly does not follow the prescription given in his notes, so perhaps they were made when he was still experimenting.

Dr Baigent died in 1935. How unfortunate it is for anglers of today that he did not live long enough to write his book.

References

1. Taverner, Eric, *Fly-tying for Trout*, London, N.D. [1942].
2. Hughes, Thomas, 'How to Breed Fowls for Fly Tying', *Fishing Gazette*, 1922.
3. Id., 'Breeding Game Fowls for Hackles for Fly Tying', *Salmon and Trout*, No. 50, Oct. 1928.
4. Id., 'Feathers for Artificial Flies', *Feathered World*. Reprinted in *Fishing Gazette*, 1928, and *Flyfishers' Journal*, Vol. 21, Summer 1932.
5. Henderson, John, 'Some Notes on Blue Game Fowls and Hackles', *Flyfishers' Journal*, Vol. 21, Spring 1932.
6. Id., 'Retrospect' (On the Breeding of Old English Game Fowls for their Hackles), *Flyfishers' Journal*, Vol. 51, Autumn 1962.
7. Id., 'The Old English Game Fowl and its Breeding for Hackles', *Fishing Gazette*, 1949.
8. Elder, H. F. D., 'The Fly Fishers' Club Collection of Fly Tying Material', *Flyfishers' Journal*, Vol. 64, Autumn 1975.
9. Id., 'Dr Baigent's Flies', *Flyfishers' Journal*, Vol. 56, Winter 1967.

CHAPTER II

Feather structure

The detailed structure of a feather is of no importance to a fly-dresser, but there are a few points which should be clarified so that the parts of the hackle referred to can be identified.

The normal body feather on a bird consists of at the base, the BARREL or *calamus*; up from it extends the SHAFT or *rhachis*, the point of which forms the tip of the feather. On each side of the shaft project the BARBS or *rami*, and each barb has growing out from it on both sides, at nearly right angles, BARBULES or *radii*. The barbules on the upper edge of the barbs have minute hooks, while those on the lower edge are hookless. The hooks on the barbules on the upper edges grip onto the barbules of the barb above and this makes the feather into a firm surface often called a web. In the flight feathers, the barbs are so firmly fixed together by the barbules, that an almost air-tight surface is formed.

In a good adult cock's hackle suitable for a dry-fly, the normal feather structure has been modified because, fortunately for us, the hackle of the wild Red Jungle Cock, the ancestor of the domestic fowl, has developed its neck hackle as a display feather; the barbules are mainly missing.

In a hackle, the barrel, which holds the hackle to the skin, is very short, being a hard thick portion about 1 mm long at the base of the shaft. The shaft extends up the centre to the tip of the hackle and in a good hackle it will be stiff and springy; it is the shaft which is wound round the hook when tying a fly. Just above the barrel there is a tiny fluffy feather attached to the back of the shaft called the aftershaft; it is of no importance to us as it is removed when the base is stripped to prepare the hackle for tying. The barbs are opposite each other on the shaft, not set alternately, and in a hackle are of equal length on both sides of the shaft, unlike some specialised feathers, such as the flight feathers, where the web on one side is broader than on the other. In a hen's hackle the barbs normally carry barbules all along their length giving a solid feather, but in a cock the barbs have mainly lost their barbules except near the shaft, each barb being separate from its neighbour. It is the strength of the individual barbs, the sheen and colour developed for display and the lack of barbules which makes the cock hackle suitable for a dry-fly. In a hen's hackle the mass of barbules – there must be many thousands on a single hackle – hold the water and make drying of the fly impossible.

The length of barb carrying barbules varies from one cock to another. In a poor dry-fly hackle, the barbules extend well up the barbs and there is a broad web on each side of the shaft, but in a good hackle the barbules only extend out for a short distance and the web is narrow. Towards the base of the hackle, the barbules always extend further out and the web here, even in a good hackle, will be quite broad. Below this, the normal barbs and barbules are replaced by what looks like down and this, along with the after-shaft, is always stripped off the shaft when preparing the hackle for tying.

Throughout this book I have used the names I have given to distinguish the parts of the hackle. Some may dislike the use of the word barb to name the part which sticks out from the shaft and does in fact do all the work in supporting the fly on the surface. It is the important part of a good dry-fly hackle. If the barbs are stiff, shiny and with few barbules, it is a good hackle. The only difficulty in using this name is that we use the name barb for the barb of the hook, but I do not think that this should cause confusion. This name is freely used in America and there is no other name which has been universally agreed in angling or fly-dressing books to name this part of the hackle, although fibre and plume are often used, the first being the commonest in present-day books. Personally I prefer barb because not only is it the normal word written in textbooks on feathers, but it easily allows the word barbule to be used. Possibly one should speak of rami and radii, but these are unknown words to the fly-tyer and we can leave them to the scientist.

It is relevant to include a short note on the normal production of hackles as the chick grows to the adult cock. On hatching, the chick is covered with down, a feather of a special character. These down feathers are soon replaced by the first normal type of feather and the hackles of the male and female in this plumage are similar. As the young bird grows, the first hackles are replaced by a second set of hackles and at this stage the difference in shape of the cock and the hen hackle becomes apparent, but the hackles of the cock do not have the real stiffness of the adult bird. The final and adult hackles come in with the full plumage which is produced when the bird is six to nine months from hatching. Unfortunately for the breeder for hackles, the cock neck hackles are almost the last feathers on the body to complete their growth. A young cock can appear to be in full plumage and yet when examined, many of the neck hackles will not have grown free from their sheaths and, as every fly-dresser knows, a hackle still with part of the sheath at the base gives a poor result.

I am not going to give an account of how a feather grows, for this is of no practical importance and anyone can find this information elsewhere if they so desire, but there are one or two points which are of general interest.

A hackle, once it has completed its growth and the growing sheath has been discarded, is no longer a live thing, in the sense that it does not continue to grow and there is no passage of blood or nutriment from the body into the feather. It is dead. It is for this reason that a hackle remains in exactly the same form as when it was on the bird. It does not wilt or turn limp or become brittle. The colour can change with age, but the texture remains the same. At the time of moult, which in the fowl is every autumn, all the hackles are replaced, the old hackle being thrown out by the new one growing in.

Some writers, who have kept cocks for hackles, refer to the practice of plucking hackles from the cock's neck twice a year and although I do not think that it specifically stated in any book the inference is that when a hackle is pulled out, the bird grows a replacement. The experts on poultry that I have consulted are, strange to say, not prepared to give a clear statement as to whether feather replacement is or is not normal in a bird. All I can say is that with my birds this does not happen. When penning my breeding cocks with their hens, I normally pluck the useful sized hackles from their necks, for the hens often pull and spoil the hackles of a cock. These hackles provide me with a useful supply for my own use. I had expected these hackles to start growing in again immediately, but I have never found this to happen; the hackle is not replaced until the following moult. Whether I have produced a rather strange race of birds or whether replacement is not in fact normal, I do not know, but it certainly does not happen with my strain.

CHAPTER III

The Jungle Cock

The evolution of the domestic fowl through the ages is of limited importance as far as the hackle is concerned. The point which is vital is that the ancestor of our domestic cock, the wild Jungle Fowl, developed the feathers of its neck into an ornamental feather used for display in courtship and aggression. The neck feather of the male has become long, pointed and looks shiny because, for the greater part of its length, it has lost its barbules, leaving only the stiff polished barbs. Fortunately the domestic cock in the majority of breeds has, during its evolution, retained this characteristic of the original wild bird, for the Jungle Cock is the only bird in the world which has developed its display plumage in this way. Had it developed ornamental feathers of another kind, dry-fly fishing as we know it would be impossible, for the only material known which can carry out the dual role of supporting the fly and at the same time producing a representation of the wings and legs of a live dun, is a hackle taken from the neck of a domestic cock. It is indeed a miracle that nature has given the angler such a perfect material to work with.

The Jungle Fowls, the genus *Gallus*, belong to the Pheasant family, which are natives of the eastern part of the world. As a family, the males tend to have very ornate plumage, while the females are rather drab. The common Pheasant and Peacock are good examples. The Jungle Fowls live in the scrub on the edge rather than in the denser wooded areas. They move about in parties and although when required their flight can be fast and direct, they tend to keep out of trouble's way by quietly disappearing into, round or through the undergrowth. The domestic fowl has inherited its ancestors' habits almost exactly. Try to find a flock of game fowl in an overgrown paddock or garden and you see the typical behaviour. You know that they are there, but you are lucky if you get more than a glimpse of one as it disappears round a bush. Like the wild bird they are most active in the early part of the day and one thing that they seem to hate is bright sunlight.

There are four species of Jungle Fowl in the world: the Red Jungle Fowl, the Grey, the Green and the Ceylon species, the last being somewhat similar to the Red in general appearance. I have not covered the full literature dealing with the evolution of the domestic fowl, but I understand that it is normally considered that the Red Jungle Fowl alone is the ancestor of our breeds. There is no evidence to suggest that any of the other species are in its ancestry, and although I am no taxonomist, judging from the hackles alone, I would support this view, since no variety of poultry has, to my knowledge, the fused tip of the Grey Jungle Fowl, no variety has a green shade in its plumage as in the Green and there is no sign of the uneven tip of the Ceylon. Delacour,[1] however, states that the Jungle Fowl in their natural habitat freely cross with the native domestic breeds and John Henderson, who spent many years in Malay, has told me that crosses with the wild bird were frequently seen in the local poultry yards and that he had shot birds in the wild which plainly showed the influence of the domestic breed. If, as Delacour states, this is also a common occurrence with the other species, it would seem reasonable that through the ages the blood of some of the others would have come into the race. I should very much like to see the results of the crossing between the Grey, Green and Ceylon with the local domestic breeds. The possibilities for the fly-dresser are fascinating. Beebe,[2] however, says that while the Red species crosses freely in the wild with domestic birds, crosses between the domestic and the Ceylon or Green are rare. He is in no doubt that the Red and the Red alone is the ancestor of the domestic hen.

The Grey or Sonnerat's Jungle Fowl (*Gallus sonnerati*) is the 'Junglefowl' of the salmon or sea trout fly. A most curious ornamental evolution has taken place in the neck hackles and to a lesser extent in the minor wing coverts of the male bird. In the neck hackle, the barbs of the feather near the tip have become fused together and form two solid shiny plates. In the adult bird the plate at the tip is a shiny yellow chrome and there is a smaller white plate about a quarter way down the hackle. In the first year bird the plates tend to be white. For the wet-fly this shiny-eyed hackle used as a 'cheek' is invaluable, but for the dry-fly fisherman the hackle of Sonnerat's Jungle Fowl has little use, the hackle below the tip not being very stiff. This species is confined to western, southern and central India, but it has become rare due to the uncontrolled slaughter for its hackles. Since 1967 efforts have been made to ban the export of skins of this species from India and it is to be hoped that this unique bird will not be exterminated. I understand that it is not particularly difficult to breed in this country, so we may see a flock of Grey Jungle Fowl being built up to supply the salmon fisherman with the hackles that he requires for many of his flies. It would appear, however, that it would not be very difficult to manufacture a substitute, which would give an equally good 'cheek', since the feather is not spread or wound round the hook and is used solely to show the enamelled plate.

As in the Red Jungle Fowl, the male of the Grey has an 'eclipse' plumage: the beautiful neck hackles are moulted at the end of the breeding season and are replaced by dull brown feathers with no white tips. After the autumn moult, the fine neck hackles appear once again. The female of the Grey is a dull plumaged bird, whose feathers are of no particular use in fly-tying.

I do not think that crossing this species with any domestic breed would serve any useful purpose. The use of this bird to the fly-dresser is to supply the white cheeks, so if it is to be bred in captivity it would be best to keep the species pure in order to supply this requirement. I cannot imagine that crossing could ever improve this quality in the hackle and there is no particular colour or other characteristic which one would want to transfer to another strain. I hope that a pure flock of this unique species will someday be built up in Britain.

The Green Jungle Fowl (*Gallus varius*). This interesting species, which is found only in Java, is very different from the others and it has evolved along different lines. The neck hackles of the male are not elongated as in the Red and Ceylon species, nor very ornamental as in the Grey. The whole bird is a dull grey-green, the neck hackles being rounded, with barbules to the tips of the barbs as in a domestic hen. The round points are edged with a dark black-green band of about an eighth of an inch and behind this there is a metallic green sheen, which reminds one of similar markings in red on the common Pheasant. This green colour is the main point of interest, for in our domestic breeds there is not one with any sign of green in its plumage. Green, or grey-green, willow as it is called, in the legs is not uncommon in some breeds, but not green in the feather. The general shape of the bird is different from our domestic hen and this along with the green colour would appear to indicate that this species also is not represented in the ancestry of our own birds. If this jungle fowl could be crossed with a game fowl some interesting birds might be produced, for if we could get green into our range of poultry colours, some fascinating possibilities come to mind.

Lafayette's Jungle Fowl (*Gallus lafayettii*). This species is confined to Ceylon. It is not unlike the Red Jungle Fowl, but the breast is a mottled brown, not a pure black; the neck hackles are more of a yellow red and in these hackles the black list is more pronounced. The hackles with this list tend to extend further up the neck than in the Red, so that it appears to a greater extent in the smaller hackles.

Delacour[1] states that the neck hackles of this species are very long, but this did not appear very evident to me in the series that I examined in the British Museum. All the ornamental hackles, neck, saddle, spade, have a curious formation which for some reason is not mentioned by Delacour[1] or Beebe.[2] The tip of every hackle has lost two or three of its barbs, which gives it a frayed look, the end of the feather ending in a V. This is clearly seen in the illustration (Plate I). From a fly-tyer's point of view, this would be of little importance, the tip being either tied in or discarded depending on how you tie in your hackle, but it is a most interesting difference from the other species.

There is also another difference in this species as compared to the Red. Delacour[1] states that the male of Lafayette's does not attain full plumage until the second year; in the first, the neck hackles are rounded and blackish-brown in colour.

It is difficult to say whether this species had any place in the ancestry of our domestic hen; certainly I have not seen any sign of the strange formation in the tip of the hackle in any of our breeds, but this is probably not conclusive. Herter[3] stated that where a cock had furnace or badger hackles with a very well pronounced list, it proved that the Ceylon species was in its ancestry, but he gives no facts to support this statement. Since the Red Jungle Fowl has a well pronounced list and since list can easily be extended into the small hackles by selective breeding, I can see no reason why this statement should be accepted. Herter said so many extraordinary things about hackles that without further evidence to support them they can be disregarded.

The Red Jungle Fowl (*Gallus gallus*). This is the most widespread and also, from the fly-dresser's point of view, the most important species. It is found from the north-west of India right through to China, down through the Malay peninsula and into many of the islands in the Malay Straits. To look at, it is very like a small black-breasted red Old English Game. About the size of a large bantam, it has a black breast, tail and wings, reddish on the back. The neck and saddle hackles are bright fiery red with a dark stripe up the centre shaft in the saddle and in the lower neck hackles. In the smaller hackles near the head, the dark shows only near the base of the hackle or is absent. This characteristic is common in our domestic birds, the dark centre mark or list fading out as the feathers get smaller, and for this reason a good furnace hackle in the smaller sizes is difficult to obtain. The call of this jungle fowl is rather different from that of the normal barnyard rooster and to my eye it has a most distinctive way of walking. At each step, the action is to lift the leg very high, close the toes together, then shoot the leg out almost parallel to the ground, spread the toes wide and bring it down. It reminds one of the trot of a hackney horse in a most deliberate slow motion. Each step is a separate clear-cut action and it looks as if it was going over an obstacle each time. No doubt the Red Jungle Fowl has developed this action in order that it may easily walk or run over the thick grass in its natural terrain and I mention this characteristic because it is one which is passed on when it is crossed with the domestic bird. It is easy to see the jungle fowl parentage showing up in the offspring when they walk.

The wonderful thing from the fly-dresser's point of view is that the Red Jungle Fowl has the perfect hackle for shape and stiffness. As will be seen from the illustration (Plate I), the shape is perfect, the barbs very short in relation to its length, and it is very stiff, with a high number of barbs in a given length. These points are most important since in animals and birds it is always easier by selective breeding to produce a strain which goes back to certain of the natural characteristics of the original ancestor rather than attempt to produce new characteristics. The one unfortunate point in the hackle of the Jungle Cock is the colour of the reverse side, which has a chalky whiteness. This makes an unattractive fly and this characteristic tends to be carried into the domestic fowl, although it can be reduced by careful breeding.

The Red Jungle Fowl according to Delacour[1] and Beebe[2] crosses freely with the other species, producing fertile hybrids, and it also crosses easily with the smaller domestic breeds. They both say that the majority of the birds seen in captivity in Europe and America are not pure, having domestic blood somewhere in their breeding. In the true breed the hen bird has no comb or lappets and the cock has a well developed eclipse plumage. The eclipse plumage is assumed by the cock at the end of the breeding season. The long pointed red neck hackles are moulted and replaced by soft rounded hackles of a blackish-brown colour (Plate I). At the full autumn moult the red hackles are regrown. In birds which are not pure the tail of both cock and hen is held upright while the wild bird holds its tail almost horizontal. The cock's crow is also not the true call of the wild bird, but the certain pointer is whether the cock assumes its eclipse plumage, for this plumage is only seen in a bird which is completely pure bred. The Red Jungle Fowl (*Gallus gallus*) has been divided into five races, the species having evolved slightly differently in the course of time in the various areas in which it lives. The races, according to Delacour,[1] are well marked and easily distinguished in life, but not widely different. As is the way in ornithology these races have been given different names.

The typical race (*Gallus gallus gallus*) extends over the area from east Thailand to Cambodia and Cochin China. The Burmese race (*G. g. spodiceus*) from Burma extends to west Thailand, Malay and down into north Sumatra. The Tonkinese race (*G. g. jabouillei*), Tonkin and Hainan. The Indian race (*G. g. murghi*), north and north-west India to Bengal. The Java race (*G. g. bankiva*), in Java. In his book Delacour[1] states that one of the main differences between the races is the shape and colour of the neck hackles. He says that of the five races the type (*G. g. gallus*) from China has the longest and most pointed neck hackles, the Burmese has shorter hackles, the Tonkinese has darker plumage and the neck hackles are shorter and a little less pointed than the type. The Indian has yellower neck hackles, the longer ones golden-yellow with a broad central stripe. This race he says is inclined to be tamer and more easily domesticated, but becomes coarse and large after several generations in captivity. It is the most frequently imported but least desirable form. The Java has broader neck hackles which are almost rounded at the tip and are of a different shape from those of the other races.

It appeared from this that for a fly-dresser there could be a great difference in the quality of the hackles according to the race and it was therefore important to examine these races from a fly-dresser's point of view. With the help of correspondents and friends in the East, I have been able to collect samples from all the species and races, but for a true comparison a large number of specimens are required. At the British Museum at Tring I was able to examine a full range of skins of all the races of the Red Jungle Fowl. I am certain that Delacour, with his trained eye, could note differences in the hackle length and shape between the races, but as far as a fly-tyer is concerned, there is no practical difference between them. We are dealing with a single hackle while a taxonomist is comparing a series of birds against each other, and a trend in hackle shape, which he can see, is not apparent when single hackles are compared against each other. According to Delacour, the hackle of the Java race should be distinctly more rounded than a hackle from a Chinese bird, but comparing single hackles against each other, I can see little difference. In colour there was quite a difference between skins and this did not necessarily depend on race. Some birds had very yellow-red hackles, while others were a dark red. There is possibly a difference between individual birds, but it would be dangerous to take the colour from skins in a museum as proof, since colour of plumage in dead birds can change with time and in particular if they are exposed to light.

It was interesting to note that in all the races the dark black-brown stripe up the shaft – the fly-tyer's 'list' – was well marked in the large hackles near the base of the neck, but this had disappeared by the time hackles of a size used for tying dry-flies had been reached. In the Indian race, the list was darker and more extensive.

To sum up, we can say that all the races of the Red Jungle Fowl have hackles which can be classed as perfect for shape and quality, that the dark list is common to all the races, but most clearly marked in the Indian, and that as far as a fly-dresser is concerned, in a single hackle, any difference in shape from different races is not apparent.

References

1. Delacour, Jean, *The Pheasants of the World*, London, 1951.
2. Beebe, William, *A Monograph of the Pheasants*, Vol. 2, 2nd edition, London, 1921.
3. Herter, George Leonard, *Professional Fly Tying . . .*, 19th edition, U.S.A., 1971.

CHAPTER IV

Dry-fly fishing – The imitation of the newly hatched dun

The flies which the trout take off the surface of the water mainly belong to the family Ephemeridae. Sedges, ants and other such flies frequently find their way on to the surface and are taken by the trout, sometimes madly as when there is a hatch of Grannom, a form of sedge, but when the trout are 'rising' it is normally to a hatch of the up-winged flies. The different species have different life histories, but in all there are four distinct phases: the egg, the nymph, the dun or subimago, and the spinner or imago, the perfect adult insect. After mating the female spinner returns to the water to lay her eggs. After a certain length of time in the water the eggs hatch and the nymph swims free. The nymphs of the various species behave in different ways, some being free-swimming, others crawling about the river bed, others never leaving the shelter of the weeds as they grow and approach the time when they make their next change of shape. In the nymph stage the insects are purely aquatic, never leaving the water, and they are freely sought after and eaten by the trout. After a varying length of life as nymphs, they must come to the surface and turn into the winged insect, the dun. This is a very difficult and dangerous operation. What triggers off the moment when the change is to take place we do not know, temperature, atmospheric pressure – it must be some such influence for it is normal for 'a hatch', as the change from nymph to dun is usually called, to take place over a stretch of river at the same time. The nymph must leave the shelter of the weeds or river bed where it has been hiding from predators and swim to the surface. There it has to force its way through the surface film, for surface tension works in both directions, the same amount of force being needed to break through to the air above as is needed to enter the water. At the moment that it has to use physical force to break the surface, it is also undertaking the amazing change from a swimming creature to a beautiful and delicate insect with wings. The nymph case has to split and the dun emerge. The wings are almost pumped up into the upright position by the blood coursing through the veins, and the dun which emerges sitting on the remains of the nymphal skin or 'shuck' moves off on the surface of the water while the discarded shuck floats away or sinks.

It is important to realise that the newly hatched dun does not 'float' on the water, which implies some part of the insect being immersed, it sits *on* the surface. The points of its six legs support the weight of the insect on the top of the surface film. Should even the tip of one of these legs break through the surface the dun may be doomed if it cannot produce the physical strength necessary to pull it back through the surface; and if the insect remains wet the water will tend to flow back over it once again. The newly hatched dun is and must remain completely dry if it is to live. Sitting on, not in, the surface of the water it sails down the river like a tiny yacht waiting for its wings to dry out when it can lift off and become airborne. The change from the dun to the spinner does not take place on the water and is therefore of no interest to the trout.

This amazing action by which a free-flying insect can split its skin down the back and emerge as even a more delicate thing of beauty is quite extraordinary. The spinners fly, mate, and the females return to the water to lay their eggs. They can return to the water in a number of different ways and it depends on their species as to which method they adopt. In actual practice it is only the species which lay their eggs on the surface of the water that are really of interest to the fisherman. In these species, and they are in the minority, the females return to the river and hovering over the surface dip and rise as they drop their eggs. Having laid their eggs their life is over and they fall dying on the surface with their wings outstretched, known to the angler as 'spinner fall'. As they lay their eggs and as they lie dying on the surface the trout can reach them and there can be great activity amongst the fish at that time. The spinners of the majority of species, however, do not lay their eggs in this way. Encased in a bubble of air and with wings folded they crawl under the water from the edge of the river or down vegetation sticking through the water and lay their eggs underneath the surface. Having completed their life cycle they die and are swept away in the current. It is for these species that fly boards are installed on the chalk streams, for they provide a platform on which the spinner can land and creep down to lay her eggs on the under surface. A rise to spinners is not really a very common occurrence in the majority of rivers; it is the spectacular rise to the spinner fall of the Mayfly or the B.W.O. that has made it so well known.

There are therefore four distinct times when the trout take the Ephemeridae in one of its forms:

The nymph under the water.
The nymph at the moment that it is changing into a dun in the surface film.
The dun as it sails down the stream waiting for its wings to dry.
The spinner of certain species as it dips to the surface to lay its eggs and as it lies dying on the surface.

The nymph lives for months and sometimes for as long as a year in the water. During the whole of this time it is preyed upon by the trout and it is in this form that by far the greatest number must be taken by the fish. Their life must be a constant game of hide and seek with the trout, culminating in the last hectic dash to the surface to hatch into a dun. In that dash, the only hope for survival is speed and the certainty that amongst the large numbers hatching at the same moment some must escape. In the nymph form the insect is completely under the water and the angler can represent it with an artificial nymph or a wet-fly. The representation is fished under the water and should penetrate through the surface film as quickly as possible.

On certain occasions the trout seem to be very keen on taking the nymph at the moment that it is breaking through the surface film and turning into a dun. This happens fairly frequently with a hatch of March Browns, when they hatch in the rough water at the head of a pool. I have seen the place where they are hatching boil with rising trout, but lower down the pool not a rise to be seen and the hatched insects sail down untouched. An imitation of the hatching dun should be fished in or just under the surface film. This calls for a sort of soggy fly quickly sinking yet sufficiently buoyant to prevent it sinking too far below the surface. A fur or herl body which can hold sufficient air to prevent quick sinking would appear to be the type of material required to represent the hatching dun.

Watching the trout take the newly hatched dun as it sails down the river is the most exciting sight for an angler and the representation of the dun and hooking trout with this artificial fly is what dry-fly fishing is all about. But let us be perfectly clear as to what we are attempting to do. We are attempting to represent the newly hatched dun as it sails down the river waiting for its wings to dry, before it lifts off the surface and flies to the comparative safety of the river bank. It is sitting *on* the surface of the water, not a single part of it is in the water. It is in fact *dry* and if we are to represent the insect in this stage of its life, our representation should also be *dry*. In fact we must fish a 'dry-fly'. It may appear that I am labouring the obvious, but there has been a lot of muddled thinking since anglers first tried to produce a representation of the dun. For a true copy, it must be a *dry* fly, it cannot be a fly floating in the surface film. The artificial must sit *on* the surface. Not the tip of a single barb should penetrate the surface if it is to be a true imitation.

Since the days of Halford, fly-tyers have produced many variations which they suggest are improvements on the original dry-fly tied with a cock's hackle. 'Parachute' flies, flies tied with two bunches of hackle spread out on either side to support the weight of the fly on the surface, flies with the bottom hackles cut away in a V and many other such ideas are examples. All these variations are supposed, by their inventor, to be a step forward in the evolution of the dry-fly. They are of course nothing of the sort, all that they are attempting to do is to produce an alternative to the proper dry-fly. The dun sits *on* the surface of the water, so the imitation must do the same. The only substance so far known which is delicate enough and strong enough to attempt to resemble the legs of a dun are the barbs of a really stiff cock's hackle. Some day a synthetic substance may be produced which can improve on this natural material, but so far there is nothing which has this strength. The trouble is that these really stiff cock's hackles are not easy to obtain; they never have been and ever since dry-fly fishing became popular, there has always been a search for good hackles.

A parachute fly or a fly tied with a bunch of hackles out on each side is not a representation of a newly hatched dun. It is simply a production which is easily seen and will float for a long period before becoming water-logged. The fact that it will catch fish and indeed at certain times may be more effective than the really good dry-fly is not the point. It may be taken by the trout as a dun at the moment of hatching or a dun which has become trapped in the surface film and if the trout at that time prefer a representation of that type then it would be stupid not to use such a fly. Only the purest of purists would refrain from doing so. But we should clearly realise that we are representing a dun in a special condition, we are not representing a dun as it sails down the stream waiting for its wings to dry.

The dun is in great peril from the trout while it is on the surface, so it is not surprising that it lifts off in flight as soon as possible. Frequently two or more attempts are made before it becomes airborne. Once in the air it heads for the shelter of the foliage on the river bank, attempting to evade the attention of the Swifts, Swallows, Chaffinches and Wagtails which always seem to collect and take their toll when a hatch is in progress. On reaching the bank, it hides until it is properly dry and able to fly freely.

The insect is now of no interest to the fisherman as it does not return to the water in the dun stage. After a certain length of time the final wonderful change takes place. The free-flying dun splits its skin down the back and the beautiful spinner, the perfect insect emerges. The spinners mate, the males die on the river bank, but the female before she can complete her life cycle must return to the water to lay her eggs. If she dips to the surface to lay her eggs she is within reach of the trout and, as with the newly hatched dun is on the surface. When she falls dead or dying, however, then we have a different subject to copy. She is now not on the surface but in it. The wings are not upright but flat on the water. For a good representation here, I feel that the body and the outspread wings are the salient points and the legs of little importance. It must really be rare for the trout to see the spinner sitting on the surface with upright wings; normally she will be seen in the 'spent' position with wings outstretched lying in the surface film. I find it strange therefore that there are representations of the spinner tied with stiff cock's hackles to keep the fly above the surface, when soft hackles or no hackles at all, to allow the fly to sink into the surface, would appear to be the best solution. This aspect has been discussed further in a later chapter.

It is easy for us to see what a dun looks like on the surface, but if we are to deceive the trout, the important thing is to attempt to understand what the insect looks like in the eyes of a fish looking upwards from underneath, not downwards from above, and we must present our imitations accordingly. Much has been written about the trout's 'window' but for some reason there has been little written explaining exactly what the trout sees from under the water. There is, however, one book, *The Flyfisher and the Trout's Point of View* by Col. E. W. Harding,[1] which never seems to have received the attention that it deserves. It was published in 1931 and Harding had made a careful study of the optics involved to give a very clear picture of the 'trout's world'. He constructed a tank which would allow him to look upwards through the water and to check his results against what the fish could actually see. Possibly this book is not well known because the author treated the subject in such detail that the explanations and diagrams are too complicated for the average fisherman. The points he makes, however, are so important to our understanding of what the world of the trout is like and have such a bearing on how we should approach the problem of producing and presenting an artificial fly in the most likely way to fool the trout, that they should be understood. I have taken from this book the aspects which I consider are fundamental to our understanding of what the fish sees when we present a dry-fly.

To a trout lying under the surface, his world is very different from ours. In Fig. 1 I have tried to show diagrammatically what happens when the trout looks round about. F is the fish and S the surface of the water. If he looks directly upwards, FOA, he can look through the surface in a straight line just as we can see in a straight line through air. If, however, he looks slightly forwards and upwards, what is known as refraction takes place, the line of sight is bent as it goes through the surface and FOB is not a straight line. Further forward still and as the line of sight passes through the surface at a more acute angle the refraction is even greater, FOC, and he is getting near the edge of his 'window'. If he looks forward at an angle of more than $48\frac{1}{2}°$ his line of sight will not pass through the surface at all. Beyond this angle the surface acts as a mirror and the line of sight is reflected downwards as in FOD, FOE and FOG. (This also works the other way round. If we look forward into the water at an angle of more than $48\frac{1}{2}°$, we cannot see the bottom of the river, we see a reflection of the sky above.) He can also of course

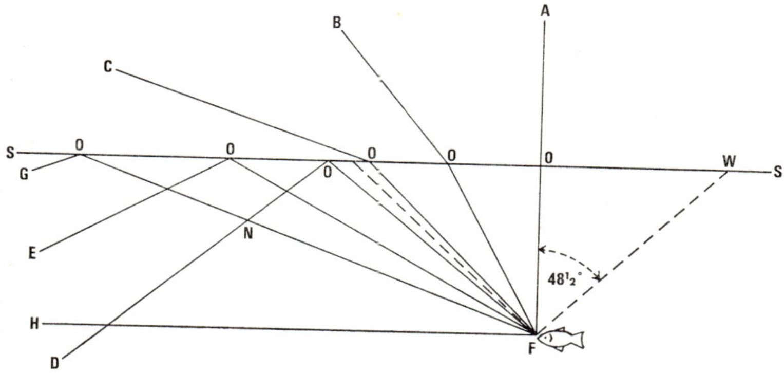

FIG. 1

look directly through the water as FH. Thus if there is a nymph at N, he can see the underside of it by looking directly at it through the water FN, or he can also see the top side of it by reflection along the line FOD. What an 'Alice Through the Looking Glass' world! A

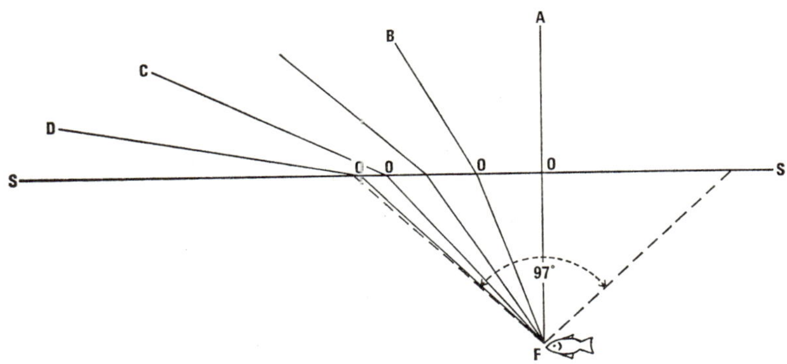

FIG. 2

circular window above his head through which he can see into the world and surrounding this a shimmering misty mirror on which by reflection he sees the bottom and anything in the water. Straight ahead and he sees things as we do.

If a trout is lying 18 inches under the water, his 'window' is a diameter of 40 inches. If he rises to 9 inches below the surface, his window is reduced to 20 inches diameter. However, it must not be thought that because he has only this small hole to look through, that he can see very little of the outside world. As I have attempted to show in Fig. 2, owing to refraction at the surface he can see a lot. As his lines of sight approach the outside of the circle, the refraction becomes very large and at the very edge it is bent so that it runs only 5° above the surface, FOD. The increasing refraction towards the edge of the 'window' does, however, produce a distorted vision. A man 6 feet tall standing 5 yards away from our eye subtends an angle of 22°, but if this man is standing just above the surface 5 yards away from our fish he subtends an angle of only 5°. Instead of 6 feet he would appear to be only 15 inches, and if he were 10 yards distant he would look 9½ inches.

This does of course work in both directions. In Fig. 3 the fish F sees the angler A along the line of sight FOA and the angler sees the fish along the same line of sight AOF, for as we look into the water our line of sight is bent in the same way whether it goes from the air into the water or vice versa. However, we don't realise that our line

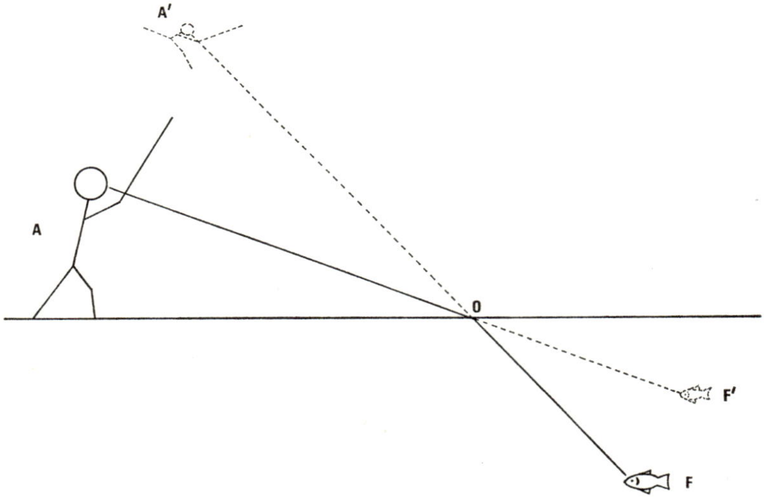

FIG. 3

of sight has been bent, we think that we are looking straight at the trout and it seems to us that the trout is lying at F′, i.e. nearer the surface and further from us than it actually is. In the same way the trout will see the man at A′, not at A, and, because of the reduction in the angle due to refraction, if the man is 5 yards away he will appear to the trout to be his correct breadth but only 15 inches tall and of this 15 inches his upper parts, being higher, will be disproportionately large. The motto is to keep low down to the water to keep hidden from the trout. When we see our fish therefore, it is nearer to us than it looks, it is deeper in the water and it can be larger than we think. It is interesting to note that a young Kingfisher out of the nest has the same difficulties to contend with. It dives straight at the minnow and misses every time. It has to learn to dive inside the point where it sees its fish.

There is another important result caused by the increasing refraction as the fish looks towards the edge of his window. The angle subtended by an object is reduced, 22° is reduced to 5°, but this also means that the amount of light in 22° is condensed into 5°. The intensity of the light round the edge of his window on a sunny day must be blinding, particularly when his eyes naturally magnify more than ours. This clearly explains why he heads for the shade in bright light. It also explains why he has the best of it when the light begins to fade or even in what is to us pitch darkness. For him there is 4½ times more light round the edge of his window and he can see perfectly clearly. At night a dark fly which to us is invisible is easily seen by the fish.

For the trout, however, his window is but a very small part of his field of view and in Harding's opinion probably not a very important part of his life. Through it he can watch for danger approaching from outside his medium, but it can, in the main, have little importance in regard to his food supply. He can only feed on objects under or on the surface. Nymphs and other food under the surface can be seen and captured by direct vision or by reflected vision from the surface film, but how does he see and rise to a dun on the surface? If a feeding trout lying 9 inches under the surface waited until he saw a fly in his window, in a slow current of 2 m.p.h., he would have 3½ seconds between first seeing it and when it was directly above his head. With a 4 m.p.h. current slightly over 1½ seconds. In this time he would have to decide whether it was good food, position himself and rise 9 inches upwards to take the fly. He could of course drift down with the current as he inspected the dun, but while a trout does occasionally do this, it is not the normal way in which he behaves. The move sideways to position himself is done as the insect approaches. The 3½ and 1½ seconds are based on the assumption that the fly is going to pass directly over the trout's position. If it were slightly to one side, the trout would have even less time, since it would be further down the stream before it came into his window and if it were more than 10 inches on either side of him, he would never see it at all.

Although the trout cannot see through the surface beyond his window, because his sight is reflected downwards, anything touching the surface produces a 'light pattern' which is very clearly seen from below and Harding points out that this is a most important point which has been mainly missed when considering a trout's feeding habits. As a dun approaches our trout, he cannot see the insect's body or wings, which are above the surface, but he can see the marks made on the surface by the tips of its legs. The six legs of the dun touching the surface film do not, as we might think, appear as six minute dark dots. They produce a 'light pattern' of six distinct spots or streaks of bright light. Hewitt[2] has pointed out that these six feet touching the surface draw up the water, by surface tension, to create tiny lenses, which magnify the light into little light blobs or lines of intensity. These streaks of light are clearly seen by the trout as he looks forward and upwards through the dark water. Every insect produces its own 'light pattern' and Harding is quite certain that it is this light pattern which the trout sees approaching down the current towards him and this is the main indication that food is on its way.

The facts produced by Harding convince me that his theories as to how a trout feeds must be correct. Although the trout will be able to see the tip of the wings of the fly before it actually comes into his window, because of refraction, the size of the wing would look minute and indeed it would only be a few inches beyond the edge of the window before it became visible at all. If we accept Harding's readings of the sequence of events as seen by the trout, the light pattern made by the feet of the dun is clearly seen by the fish and he can see this pattern on the underside of the surface for a long way ahead or on either side. It has nothing to do with his window. Seeing the light pattern approaching, he can move to left or right so that it will pass directly over his head and he can also rise up in the water ready to take the dun if he decides to do so. He can move forwards and upwards to take the dun before it reaches his window or, particularly in a slow current, once he is in position, he has time to let the dun come into his window and decide whether to take it or not after visual inspection. Harding suggests that in the majority of occasions, particularly in a fast current, the trout rises to the light pattern alone, taking the fly before it reaches his window. If this is correct, it explains why at times the colour of the fly does not appear to matter, while at others the trout are very fussy, demanding an exact representation of the fly on which they are feeding. In the first case, they are rising to the light pattern produced by the approaching insect and if our fly gives the correct light pattern, colour does not matter, because he never sees it. In the second, the trout is waiting until the dun comes into his sight before he rises and if the representation is not right, he will reject it. Whether he is feeding in the first or the second manner, the essential thing for us is to ensure that our fly is producing the correct light pattern. If he is rising to the light pattern only, then this is all we require. If he is looking at the fly, the correct light pattern is still essential otherwise he will reject it even before it has reached his window.

Harding also explains why 'drag' is so fatal. Seen from under the surface a dragging fly produces a V-shaped pattern of brilliant lines of light, enough to make any trout fly for shelter, but it also explains why on occasions when the sedge is out, a dragging fly is taken. The sedge as it flutters *across* the current produces a rather similar light pattern. The first essential therefore for a dry-fly representing the dun is to ensure that the fly produces the correct light pattern on the underside of the surface. Harding says that from his own observations in his tank a really stiff cock's hackle produces a series of sharp points of light very similar to the natural dun. A soft hackle does not produce the correct sharp pattern.

There is a further explanation as to why, when the trout are feeding on the fallen spinners, they are so difficult to take. The spinner is lying *in* not *on* the surface film and the whole insect, its shape and colour, is seen by the fish. The veiled light from above shines through the body and wings showing the exact colour and luminosity of the insect. He sees all this as the insect approaches in the current and has plenty of time to inspect it and decide whether it is good food or better left alone. It is difficult to produce an artificial containing a steel hook which will stand up to this inspection.

When considering the hackles we must use to represent the insect in these various stages of its life, there are two main categories. Are we representing the insect when it is above the water or in the water? When the fly is in the nymph stage under the water, our artificial fly will be under the water, and we will want our hackle to move in the water like the legs of a nymph. Very few barbs will be required and they should move in the current in a life-like manner. It is normal to use a hen's hackle for wet-fly since the soft barbs will wave naturally and the barbules holding the water will allow the fly to enter the water immediately at each cast. Cutcliffe,[3] however, whose writing indicates that he probably had a better 'feel' for hackles than any other writer, states that he considers that a good cock's hackle is far superior to a hen's for wet-fly fishing on his fast Devonshire streams. He thought that a hen's hackle in fast water became wrapped round the hook and looked like a wet mass of dirt. He went to great lengths in order to obtain the colour and springiness he required for his hackles, for he believed that the hackle was the most important part of his wet-flies. This is not the normal thought for wet-flies at the present time; a hen's hackle is the tie usually specified.

There seem to be very few attempts to reproduce the dun at the moment of hatching probably because there is nothing very definite in shape to be copied. The best known example is the Gold Ribbed Hare's Ear which is taken on all types of stream. To our eyes it does not look like a good representation of any particular insect, but it hooks a lot of trout when the fish are feeding on the Medium Olive and it is generally accepted that it is probably taken by the trout as an Olive at the moment of hatching. When they show a preference for the hatching insect the fly should be fished in the surface film.

For the representation of the dun sitting on the surface, the hackle is the all-important part of the fly. A really first-class hackle is quite essential for a good dry-fly, for it must hold the body above the surface and therefore be invisible to the trout. The hackle is asked to do an awful lot. It must have barbs stiff enough to support a steel hook, when only the very tips of the barbs are touching the surface of the water and the hackle must be of a steeliness which allows the water to be thrown off during the false casting, so that the fly is completely dry once again when it is dropped on the water at the next cast. The only feather in the world which has these qualities is a hackle taken from the neck of a domestic cock. Just any cock will not do, for the quality of a cock's hackle varies greatly.

If we accept these arguments then in a 'hackled' fly we have two aspects to consider. First to produce a fly where the density of the hackle above the hook imitates the wings of the natural and is sufficiently life-like to satisfy the trout when it inspects it through its window. The second and surely the more important, is to dress a fly where the 'light pattern' produced on the undersurface as seen by the trout from below is a good imitation of the pattern produced by the live insect. For the first, I would have considered that a fairly bushy hackle would be required to represent the solid although moderately transparent wings of the dun. To satisfy the second requirement is more difficult. The live dun has six feet which touch the water and they produce the streaks of light which are seen by the trout. I am sure that a trout cannot count, but it is obviously important to produce an artificial fly which is supported on the surface by the number of hackle tips which are somewhere in the region of this number. If there are too many an unnaturally solid light pattern will be produced, too few and the light pattern will also not be correct. Because there should only be about six hackle points to do the work, one immediately considers that if it is to be a good imitation in this respect, an artificial must be very lightly dressed with very few turns of hackle. Let us examine this question in greater detail.

A normal hackle suitable for a 16 size dry-fly is approximately 4·5 cm ($1\frac{3}{4}$ in) in length. We remove the soft barbs at the base, which reduces the useful length to about 3·6 cm. Whether you tie in your hackle by the point or by the base does not matter; in either method you do not use about 0·6 cm of the barbs at the tip. The length actually used is thus further reduced to about 3 cm. At 20 barbs/cm you have 60 barbs on each side of the stem. How many of the barbs on the underside of the hackle are of use in supporting the fly? I have found it extremely difficult to produce a clear answer on this point. Obviously a certain number are wound round with the shaft and never stick out; some stick out but coming from the underside they are shorter and they do not therefore reach the circumference of the circle formed by the hackle tips; they cannot therefore touch the surface. I have taken two similar hackles, stripped the barbs off one side of one of them, wound them on hooks and then counted the number of barbs on each fly. Using two turns of hackle gave me a count of 24 'effective' barbs on the complete hackle as against 16 on the single-sided hackle. This means that half of the barbs on the underside of a hackle are wound round the shaft or do not stick out to their full length. For the 'effective' barbs I counted only those which reached the outside of the circumference. This figure if applied to our hackle for our size 16 hook gives us a total of 60 barbs from the upper side plus 30 from the under. In addition, as I have discussed in the chapter on Shape, since there is no such thing as the 'perfect' hackle, the barbs are not all of exactly the same length and we must reduce our number by at least 25% to allow for this. If we bring the thread through the hackle to tie off, no matter how carefully we do this, we bind in a number of barbs, which again reduces the number that are effective for supporting our fly. All these combined mean that our hackle of 4·5 cm when wound on to the hook has on average not more than 65 barbs of the correct length. In fact as soon as we start fishing the casting and false casting is a great strain on each individual barb; they are slowly broken or snapped off and this is clearly seen if you compare a new and a used fly, the new always being bushier than the old. However, let us consider a new fly tied with a single 4·5 cm hackle having 65 barbs of the correct length. These 65 barbs are spread round a circle of 360° – it does not matter how many turns of hackle are taken round the shank, the answer is the same. This is about one barb per six degrees. If you lay a fly on a flat table you will see that the barbs on approximately 30° of the bottom of the circumference touch the surface and it is only these that support the weight of the fly. If therefore there is one barb per 6°, there are 5 in 30°. Thus a normal hackle of a total length of 4·5 cm dressed on the fly will only have 5 hackle tips which touch the water. I think that most fly dressers will be surprised at the result of this analysis. The fly looks bushy and yet the number of hackle tips which do the work is small. My figures are based on the hackle being of high quality with very stiff barbs and also on the hackle being dry. If the hackle is soft, the barbs bend, a much greater quadrant of the circumference is used and therefore a larger number of barbs touch the surface. The same result is obtained if the hackle is wet as this allows the fly to sink into the surface film.

If we accept the argument that in a great number of cases, if not the majority, the trout rises to the light pattern created by the feet of the dun, how do we explain the certain fact that some species of fly are liked by the trout and others disliked? As examples, the Iron Blue Dun is a great favourite with the trout, while the Yellow May Dun is never taken. I have never seen a trout rise to a Yellow May Dun and I have never heard of anyone who has seen this fly accepted. It is obvious that the much smaller Iron Blue will produce a different light pattern from the larger Yellow May, but I find it hard to believe that the pattern of the Yellow May and that of the Large Dark Olive or Blue Winged Olive, which are freely taken, can be very different. Harris[4] suggests that the Yellow May is rarely taken because it never hatches out in sufficiently large numbers to interest the trout and that the same thing happens when there is a very sparse hatch of any species. I will agree that there has to be a good hatch before the fish rise in numbers, but even when there is just a trickle or even a single dun there is sometimes a trout that is tempted. If it were not so, it would be a waste of time to fish until we saw a reasonable hatch and rise. As it is, we all cast away and occasionally find a trout that is prepared to feed. Why does such a trout never take a Yellow May Dun? Even if we were to accept that these lone trout always wait until the fly is in their window before rising and that when they then see that it is a Yellow May Dun, which they know to be unpalatable, they refuse it, this does not answer the question, for how does the trout know that it is unpalatable? It can only know that it has a nasty taste if it has tried it. We should therefore see small trout taking Yellow May Duns, as they do other duns, for it is only in this way that they can learn not to do it again; but this does not happen, as even tiddlers leave this dun alone. I do not believe in an inherited instinct that a Yellow May Dun has a nasty taste.

If we accept that the lesson of palatability and unpalatability must be learnt by trial and error and that the fly is never taken off the surface, then the answer must be that the lesson is learnt under the surface. The trout must learn that it is not good to eat by taking the nymph on its way to the surface. Supposition I agree, but is there any other explanation?

This then leads me on to something which I have never seen discussed. When considering the rise of a trout, is there a direct connection between the nymph swimming to the surface to hatch and the dun on the surface? What starts a good rise and why does it end? To suggest that it is only the sight of the duns floating on the surface overhead which makes a trout rise does not fit the facts of the case. Take an example that we have all experienced. You are standing beside a good trout run on a good trout river and not a trout is breaking the surface. Suddenly in the stream you see a rise followed by another and another and you then realise that a stream of duns has begun to appear on the surface. The river is full of feeding fish. It can be the other way round, you suddenly see a dun and within moments more and more and then the splash of rising trout. After a time, the reverse can happen: the duns are coming down as freely as ever, but suddenly the fish stop feeding and a short time after the duns become fewer and the hatch is finished. If it was the light pattern or the sight of the duns passing overhead that excited the trout, then surely the hatch would have to be well on its way before there would be a sufficient number to interest the fish, but this does not happen, the fish are rising almost before we realise that the duns are there. Similarly once they have started, you would expect the trout to pick off every single dun until the last had disappeared, but they often stop while the duns are still sailing down the river. They cannot all have had enough at the same moment. The answer surely must be that it is the underwater activity of the nymphs rising to the surface to hatch which triggers off the desire for the trout to feed and the moment the underwater activity ceases, the stimulus has gone even although the duns are still there.

We have, on the other hand, also experienced the maddening situation where duns are streaming down the river and not a trout moves. Skues[5] attempts to explain this by saying that it is due to the fact that the trout have had plenty of underwater food and they are not hungry, but I am sure that it cannot be as simple as this. At times I have followed the stream of duns up the river and have come on a place where the trout are feeding. Can it be that it is only at this place that the duns are actually hatching and it is therefore only at this place that the trout have the correct stimulus? At other times, however, the duns appear to be ignored everywhere and in such cases, I have no explanation for the trout's behaviour, unless we fall back on Skues' 'not hungry', but is it reasonable to assume that every trout in the pool should have had enough to eat at the same moment? There must be a better explanation. I have not really answered any of my own questions, but I do suggest that the rise of a trout to a dun is not simply a business of the fish seeing food passing over his head and taking it. This may be the case when a single fish is seen to rise, but I am sure that a general rise of trout is not induced by a large number of duns on the top of the water. It is the result of some much more complex process. At some time or other when not a trout is to be seen, we have all settled down to cast and cast again over the same spot hoping to make the trout believe that there is a hatch, but we never succeed. This in itself really proves my point.

Why do we tie a fly with whisks? The setae on the live insect are always carried pointing at an angle upwards and Harris[4] has shown that if the setae do touch the water, the dun will be drowned. If the trout is waiting until the dun is in its window, I find it hard to believe that it will refuse your fly if it has not got whisks, but if you tie your fly so that the whisks will touch the water, thus helping to keep the artificial on the surface, you will be producing an unnatural light pattern which must be important. Whisks if set at a jaunty upwards slope look very attractive and are certainly essential for a professional fly-tyer and the shop window, but I cannot see that they are worth the extra labour involved for the amateur producing his own flies to catch fish. If you take the suggestion seriously, which I do not, that stiff whisks can prevent the fly from entering the trout's mouth, then you have an additional reason for omitting them. Baigent, the practical, never put whisks on his flies.

In 1921, Skues[5] wrote *The Way of a Trout with a Fly* . . . which dealt with the problem that I have been discussing; the food of the trout, why and how it feeds on flies and nymphs on and under the surface. He had followed the work of Dr Ward[6] and had himself looked up through Ward's tank at natural and artificial flies. Based on the knowledge available to him at that time, his work is a very clear account of the probable reasons why a trout takes an artificial fly. The discussion and reasoning is not taken from the work of any previous author, but is based entirely on careful personal observation.

In 1931, however, Harding produced his book giving the results of his experiments – a book with which Skues was not only closely associated, but his help and encouragement is acknowledged by Harding. Harding died in 1934 and in 1935 Skues published a third edition of *The Way of a Trout* . . . Skues adds a note to this edition saying that since 1931 Harding had done a lot of further work and 'he told me he had found solutions which he was hoping to give to the public'; unfortunately his material was not in a form which any editor could piece together. I cannot understand how Skues in 1935 could allow the third edition of his book to be published. Harding's work of 1931 produced facts about the trout's view from under the water which were quite unknown to Skues in 1921 and it appears to me to be really quite remarkable that Skues should allow the third edition of his book to be published with few if any alterations. With this new knowledge now available surely large parts required to be completely re-written! Stranger still, in his Note to the third edition where he refers to Harding's work, he says, 'It was some comfort to me that he confessed in a letter written not long before his death in August, 1934, that in no case did any of his discoveries do other than confirm the tentative solutions in *The Way of a Trout with a Fly*.'

In the literature written over the last hundred years, every writer bemoans the difficulty of obtaining really good cock hackles, particularly in the dun colours, which give the best representation of the insect wings, and all speak as if good hackles were more plentiful in earlier years. Modern books keep repeating that good hackles began to get scarce after cock fighting was made illegal since the Old English Game cock produced the finest and stiffest hackles. I have begun to wonder whether it really is true. There are very few illustrations dating from before the 1914 War which give a clear picture of what fishers of these days called a good dry-fly hackle and some pictures depict hackles which we would now consider to be of very poor quality. Halford[7] himself says that 'For floating flies cock hackles are so immeasurably superior . . . that, excepting in the case of honey duns and other very rare colours, I have *practically discarded* all hen hackles from my collection' (the italics are mine). I think that it is fairly obvious from the literature that while they liked a stiff hackle their standards were possibly not as high as those set by the dedicated dry-fly man of today. Good hackles are still and always will be difficult to lay one's hands on, but I go the length of saying that as far as quality is concerned it is probably easier to obtain a good cape today than it was a hundred years ago. In fact with the wholesale importation of capes from the East, I believe that our average quality is indeed higher than in the early days. As I will enlarge upon in a later chapter, the domestic fowl from the East has, on the whole, a very much higher standard of hackle than our modern breeds. The hackle taken from a strain of Old English Game which had been naturally selected over the years by its success in the cock pit may have been of superb quality, but I rather doubt it. Cock fighting in this country is not unfortunately dead and I have examined birds which appear on the exhibition bench, but which I know are also kept for other purposes. Many of these have first-class hackles, but not something to rave about.

On the question of colour, however, I am sure that the earlier anglers had the advantage. Fowls of all colours were still seen in every farmyard and although blues were rare, as they are today, they were still obtainable. In the early books such remarks as 'obtain one or two blue hens' are found. Today it is practically impossible to obtain a bird in this colour and it is also a very rare colour in imported capes. Ties invented in the early days frequently specify a blue hackle, which in itself would indicate that this colour must have been reasonably common at that time.

This brings us to the three factors which are normally used to describe and to gauge the standard of a hackle for use in a dry-fly: colour, quality, shape. Which is the most important of these three is a matter for the individual. Must it be of the exact shade or will something near the correct colour be good enough? Can a softer hackle be used or will nothing but the very best be good enough? Is it very important to have a long narrow hackle with short barbs or will a shorter and broader hackle do the job? These are questions which each individual dresser must decide for himself and the answer must vary according to the type of fly being tied, but in my opinion for real pleasure in dry-fly fishing a stiff steely hackle is the all-important factor. Others may not agree.

References

1. Harding, E. W., *The Flyfisher and the Trout's Point of View*, London, 1931.
2. Hewitt, E. R., *Secrets of the Salmon*, 2nd edition, New York and London, 1925.
3. Cutcliffe, H. C., *The Art of Trout Fishing on Rapid Streams*, South Molton, 1863.
4. Harris, J. R., *An Angler's Entomology*, 2nd edition, London, 1956.
5. Skues, G. E. M., *The Way of a Trout with a Fly . . .*, London, 1921.
6. Ward, Dr Francis, *Marvels of Fish Life*, London, 1912.
7. Halford, Frederic M., *Floating Flies and How to Dress Them*, New York, 1886.

CHAPTER V

Colour

It is more than time that the naming of hackles by their colour was put on a proper basis and an agreed nomenclature produced. This has never been done by an accepted authority with the result that all kinds of names have been commonly used to describe hackle colours. The same colour goes under many names and, what is even more confusing, people do not agree on what colour is indicated by a certain name.

This disagreement on nomenclature seems to be a problem that has always been with us, for in the *Fishing Gazette* of 1888 there is an article asking what is the correct colour of a honey dun. Capt. Lambert and F. M. Halford, being considered experts on the subject, were asked to give their definition, so even in these days there must have been differences in opinion in the fishing world. Thomas Hughes, who for over forty years bred Old English Game fowl to provide hackles, remarked in one of his articles about the variation of opinion on certain colours. As an example he said that a cock which one man called a perfect brassy blue dun was called a poor colour by another. In America, Reuben Cross[1] wished that there was a standard nomenclature.

The only detailed and acceptable classification of hackles which I have seen attempted was published by C. F. Walker[2] and it is a very sound basis on which to work. After the experience I have had in working with game fowl and studying hackles, I do not think that the subject need be a difficult one. I see no reason why a fairly simple classified list should not be produced which would cover all the main colours required by fly-dressers. I do not suppose that such a list will put an end to the matter and that all kinds of names will continue to be used to describe hackles, but I have taken some care in compiling this list and have chosen either the names which appear to be most commonly used or, where a number of names for the same colour are in common use, the name which I consider was the one originally used by fly-dressers.

It is not commonly realised how closely in some cases the names used for hackles come from the names used to describe certain varieties of game cocks. When, due to the upsurge in the interest in fishing, the tying of flies became much more of a business and as fishing and fly-dressing began to be considered to be a science rather than a hobby, there must have been a large number of game cocks in existence. Some were no doubt kept for under-cover fighting, but these were also the days when exhibition of poultry was a popular pastime. For one purpose or another, there must have been many Old English Game cocks about. In addition, unlike today, every farmyard and cottage in the country had its flock of hens kept for eggs and table. Almost all the cocks which strutted about in these mixed flocks would not be pure carefully selected strains for egg or broiler production as is the case today, but would be pure or nearly pure game cocks since the Old English Game was famed for the quality of its flesh, and there would still be a large stock of game left after the prohibition of cock fighting in 1849. There is an extensive and most involved series of names used by the lovers of the fighting cock to classify their birds according to the colour of the plumage. These delightful names such as 'Black-breasted Red', 'Blue-breasted Blue Dun', 'Polecat', 'Black-breasted Birchen Duckwing' to name a few, are used to describe the cock exactly and were the titles used by the 'Cockers' as those who kept cocks for the pit were called. Up till the early part of the nineteenth century there would be relatively few breeds of poultry available in Britain. The birds kept for the pit and the barnyard rooster would be in the majority, and the rooster would not be far away from the pure game. It is understandable therefore that the hackle produced by a cock of named colour should itself be called by that name.

When we come to classify hackles it is relevant that the number of true colours which appear in the plumage of the domestic fowl is very limited. The only real colours found in hackles are black, white and red, and even in these it is not easy to find a pure example. A pure white and a nearly pure black can be found, but there is not a pure red as in the Ibis. The red appears as shades of brown ranging from a pale ginger at the one end of the scale to a very dark red at the other. There is no vivid yellow as in the wing of the Goldfinch;

pure green and pure blue as in the Budgerigar are unknown.

Although it is not of great importance to the fly-tyer, who is only really interested in the colour as it appears to the eye, the green and blue in a bird's plumage is not caused by pigments of these colours being present in the cells. Green actually has yellow pigments in the cells and a light refraction off the feather gives the appearance of green to the eye. In the same way, blue is a grey feather showing as blue. A further extension of the principle gives the metallic colour of the Kingfisher which can show vivid green or blue according to the angle of reflection of the light.

The blue that we get in poultry is a very bad blue. Compared with the fine clear blues of the Budgerigar, it is nothing but a grey with a suspicion of blue through it and it is perhaps best compared to the 'blue' of steel. In actual fact, this colour is perfect from a fly-tyer's point of view, since a pure blue would be of little use in a dry-fly. What is required is a blue-grey which accurately represents the natural colour of the wings of the Ephemeridae duns. When speaking of hackles, we do not speak about a blue, we normally use the term Blue Dun and in this we are being descriptively correct. The feather when examined appears to have a dun shade through it. In some the dun is very weak appearing only as a suspicion of brown, in others the brown is very clear.

Dun in itself is a difficult colour to describe. It is a common colour in nature – the dun-coloured horse – and in certain breeds of cattle dun is almost the normal colour. It is a sort of grey-brown and some would say that it often had a pinkish tinge in it. The comparison of blue dun to the colour of steel is a good one since a very pale blue dun can be very near white like steel newly filed or cut, while a very dark blue dun can be practically black like steel long exposed to the atmosphere. In the middle of the range, the hackle can be matched with various shades of gun metal. The similarity continues further, for the amount of rust ingrained into the steel can also be compared to the amount of dun overlay on the hackle. I purposely say ingrained into the steel, since a freshly rusted steel appears to have rust on the surface not suffused through the metal. The effect of fresh rust also appears in a blue dun hackle, but in this case the red colour looks as if it had been applied from a pepper-pot on top of the blue and dun. A hackle so coloured is called a rusty or brassy blue dun; frequently the name is shortened to rusty dun or brassy dun. These very beautiful and much sought after hackles are described in detail later.

Hackles can be obtained which show almost every imaginable variation of colour and it is impossible to attempt to produce a nomenclature which would allow every hackle to be placed in a named category. To produce this would mean a ridiculously long and involved list and is indeed quite unnecessary. I have attempted to produce a nomenclature which will allow 90–95% of hackles to be classified. The other 5–10% can be treated as 'off colours' and described individually if necessary.

It is worth considering for a moment why it is necessary to produce a nomenclature at all. There is only one real reason, which is, so that the hackle used in a particular tie can be quickly and accurately identified and thus allow the same fly to be produced at a later date as the original dresser intended. If the fly is frequently reproduced, it means that it is a successful pattern and if it is to be frequently reproduced it must be tied with a hackle which is in reasonable supply. A fly tied with an unusual hackle can never be popular since it is impossible to obtain rare colours in quantity. Worse still, the fly, if it is copied, will be turned out with a very different hackle from that which the inventor intended. For practical purposes therefore, it is only necessary to be able to name the hackles which are commonly found in poultry; any other types can be described in detail.

Commander Walker[2] placed hackles under the headings of single colours, bi-colours and duns. The committee of the Flyfishers' Club[3] when classifying their wonderful collection of hackles used a rather different method. They classified them by colour from black through the reds and gingers to white and then from the pale to the dark blues. The bi-colours were not treated as a separate class. The answer is that there is no perfect method since we are dealing with two different things, colour and marking. I prefer Commander Walker's method, but I have extended the headings to single colours, bi-colours, barred, duns, and off-colours.

Single colours. This covers the hackles which are of one colour. However, except for a pure white and a pure black, a hackle is never the same shade of colour all over. Nature never colours as we do when we use a pot of paint to produce a flat colour without variation. A green leaf from a distance looks one shade of green, but examine it carefully and one sees that there is a marked change of colour from one area of surface to another and it is the same with hackles. A hackle may at first glance be a medium red all over, but on closer inspection, it is seen that the colour at the base is almost always of a darker shade than the tip and the colour near the shaft darker than the edges. On examination too, it often looks as if there are really two shades one on top of the other, a basic colour and a stronger shade overlaid. This is clearly seen in the gingers where it is rare to obtain one without a much darker overlay.

Bi-colours. These, as the name implies, are hackles of two colours, but, by common usage, to come into this class the two colours must occupy a certain place in the feather to produce a prescribed pattern. The list must be of one colour and the rest of the feather another. In a good bi-colour, the demarcation between the two colours should be clean-cut; they should not run into each other. The list is the name used by fly-tyers for the centre portion of the feather running up each side of the shaft. A hackle with this marking is called a pencilled hackle by the exhibitors of poultry and it is quite a common form of pattern in many breeds. The lower neck hackles of the wild Red Jungle Cock are heavily pencilled. The list is normally of a darker colour than the outer part of the barbs, a badger hackle having a black list with white outside and a furnace a black list with red outer.

When a bi-colour is wound on to the hook it gives the impression of a fly with a dark colour to the thorax while the outer colour represents that of the upper wings and lower parts of the legs. In theory the perfect bi-colour should have a list that was the same width from the base to the tip of the hackle. Unfortunately this is not possible, it is always broader at the base than at the tip, but a good example should not be too broad at the base and should not taper very quickly. A very common fault in bi-colour capes is to find that the large hackles at the bottom of the cape are well marked with the list running right to the tip of the feather, but as one goes up the cape towards the head of the bird, the list becomes shorter and shorter; near the top of the cape the list will not extend more than half way up the hackle, while the small hackles at the crown will have only a dark base and no list at all. In the collection of the Flyfishers' Club this pattern, which is very common in the reds, has been put into a special category named 'Black-butted

Reds'. The well marked list is a characteristic that can be obtained by careful selective breeding, and a strain can be built up which produces a good clear list right to the tip of even the smallest hackle.

It is not generally appreciated that the change in colour from the list to the outside takes place at exactly the point where the barbs of the cock's hackle cease to have barbules. I do not know the reason for this, but if a good bi-colour such as a furnace is examined under a lens, it will be seen quite clearly that the shaft is black, all the barbules are black as are the barbs themselves up to the point where they have barbules. Beyond this point the barbs are red.

The perfect dry-fly hackle should have no barbules which can hold the water and the perfect bi-colour should therefore be formed by a change in the colour of the barb itself. In theory, this should not be impossible since the bars in a cuckoo or creel hackle are caused by a change in colour of the barb itself. This is a line of breeding which may be explored some day, but until this perfect bi-colour appears, we try to obtain one with a long but narrow list and therefore few barbules.

Barred. In this hackle the pigmentation runs in bars across the hackle. In a good example the dark bars should be uniform, clean-cut without the colours running into each other, and the bars of dark and light colour should be fairly narrow and approximately the same width. A common fault in this marking is to find the dark bars indefinite, the edge of the colour being blurred. This gives a pale colour to the whole hackle.

Duns. Hackles which contain this colour are the most sought after, since they give the best representation of the wings of the up-winged duns. Although the blue dun should possibly be classified with the single colours and the honey dun with the bi-colours, because of their importance it is more convenient to group all hackles containing this colour in a class by themselves.

Off-colours. Frequently capes will be handled with hackles which cannot be placed in any of the four categories dealt with. These capes are often particularly beautiful and produce hackles which are exceedingly useful to the fly-tyer who is attempting a very exact representation of an insect, but since they cannot be repeated they cannot become part of a well known tie.

If these five headings are used, for practical purposes every hackle can be classified under one of them. I therefore propose the following nomenclature for hackles:

Single colours

The natural colours in which hackles can be obtained run from white through cream, ginger, red to black. Since one colour runs into the next as it gets darker, it does of course become a matter of opinion where one colour ends and the next begins.

1. *White*: Pure white. A very common colour in modern poultry kept for egg production. It is rare to find a high quality hackle in this colour, all tending to be of poor shape. Although this colour is not very frequently used itself, it is in demand for dyeing to produce olives and for 'photo-dyeing' to produce duns.

2. *Cream*: Ranging from off-white through cream to pale ginger. As with white, not a colour that is very frequently required. Good quality creams are easier to obtain than whites.

3. *Ginger*: This covers the range of colour from dark cream to pale red. The gingers tend to have a darker brown at the base and up the list, giving a two-tone hackle, and black at the base is also frequently found. The gingers can often be obtained in a very good shape and quality and they are a most useful colour. This colour is often called honey, but I consider that ginger is the name which is more normally used and that it gives a truer picture of this shade. I prefer to use ginger, retaining honey as the name to describe the outer colour of the well known honey dun.

4. *Red*: The red in poultry is a yellow-red in its paler shades and it ranges from a dark ginger through brown red and blue red to a very dark black red. At the end of the chapter I have mentioned how hackles often have a poor chalky colour on the reverse side. This is unfortunately very common in the reds. Game Red is a name used to describe the bright red which is the colour of the hackles of the wild Red Jungle Cock and is also the colour of the red Old English Game. It is a much stronger colour than the rather pale orange red of the commercial brown fowl.

Dark Red, Blood Red: Commander Walker used the term Blood Red and described it as the colour of dried blood. The dark reds have also been described as mahogany and purplish-red. It is a matter of opinion whether the reds should be split into two or more categories, but I find it extremely difficult to decide or describe where red should end and dark red begin. I therefore prefer to place all the reds in one category and by using the prefix pale, medium and dark one can describe accurately enough the shade required for a particular tie.

5. *Black*: Jet black. It is not easy to obtain a hackle of a true black. Frequently the colour on the reverse side is chalky and this gives the whole hackle a dull look. A really good black hackle always has a tinge of red through it which gives it 'life'. A good black cock has almost certainly a red not far back in its parentage and it is this that gives it the sparkle. A hackle dyed black lacks this natural sheen which makes all the difference between a dead and a live colour. Unfortunately blacks always seem to be soft.

Bi-colour

Hackles of two colours, the first and normally the darker colour, usually black, running in a stripe from the base up either side of the shaft to form the 'list', the second colour continuing from the list to the edge of the hackle. Not infrequently the tips of the barbs are also black giving a dark edge to the hackle. In a good bi-colour the two colours should be clean-cut, not running into each other, and the list should run right to the tip of even the smallest hackle.

6. *Badger*: Named after the colour of the hair of the Badger – black and white. A good badger should have a jet black list with white or cream outside. (The hair on the animal is actually the other way round, white inside and black beyond.) With Old English Game it is rather rare to have a pure white outer colour; cream is more common. I noted[3] that while the Collection of the Flyfishers' Club also had few examples of a pure white badger, cream was the colour of the majority, for no doubt most of these samples would come from game or near game. Not infrequently the tips of the barbs are black. A poor badger has a list of a pale grey and the colour can be broken

instead of being continuous right up the hackle. Hackles of high quality and good shape can often be obtained in this colour.

White Furnace and White Coch y bonddu: Kelson[4] was the first to produce these names. He describes a furnace as a red hackle with a black list and a coch y bonddu as a furnace but with black tips to the barbs. He extends this naming to cover a white hackle with a black list as white furnace and white coch y bonddu if the edge is also black. With the well-known and descriptive name of badger available I can see little point in using either of these names. I cannot remember any fly-dressing which specifies that the badger hackle used should have black tips to the barbs, although there may be one, and I can see no useful reason for including in a nomenclature a category to cover this type of hackle. If required it is quite simple to specify 'badger hackle with black tips'.

7. *Yellow Badger*: A badger but the outer colour yellow to ginger instead of white or cream. I consider that a pale yellow badger used instead of badger gives, in most cases, a more attractive fly. It could be said that there is no need to have this category, but as the bi-colours range from badger-white to furnace-red, it is too wide a spread to cover with only two classifications. It would be reasonable to suggest the name ginger badger, but I have never seen it used. Commander Walker gave the name yellow badger to this hackle and it is a good description, for the colour is usually more yellow than ginger. These hackles are often the finest for quality and shape.

In Old English Game the finest yellow badgers normally come from blue-breasted birds and the list, although it looks black, is in fact a very, very dark blue. If you have the whole cape, you can usually notice this by examining the large hackles at the base. In a blue-bred bird the list on these large hackles is always paler and can be seen to be blue. The smaller hackles near the head look jet black. The cape is indeed a very, very dark honey dun, but because the list is as good as black it has to be classified as a yellow badger. This is normally of academic interest, except that if the hackles from such a blue-bred yellow badger are kept for many years, the effects of light and time turn the dark-blue list brown and you end up with a rather nasty looking hackle. This is noticeable in certain specimens in the Flyfishers' Club Collection in the badger and the furnace section.[3] In the days when the Committee compiled this collection the hackles would appear to have a pure black list, but since they came from blue-bred birds, age has changed their colour.

8. *Furnace*: At the present time it is generally agreed that a furnace hackle is one with a black list and the outer colour red. As with a badger, a good example is rare. A hackle which has the black extending only a short way up the shaft or even showing only at the base is very common, i.e. the 'Black-Butted Reds' of the Flyfishers' Collection. The wild Red Jungle Cock, the Indian sub-species in particular, has a well defined black list in the large neck hackles, but it disappears by the time the head is reached. Reasonably good quality and shape can often be obtained in this colour although it is difficult to produce one of the quality of a fine red. Unlike the badgers, where a poorly coloured list is often a fault, in the furnace, the list is almost always a good dark black. The outer colour, in the game breeds, is the bright shade of the true Red Game, but it can be obtained in the whole red range from dark ginger to very dark red.

At the present day it is generally understood amongst fly-dressers that a furnace hackle is one with a black list and red outside, while a coch y bonddu has a black list, red outside and the tips of the barbs are black; i.e. furnace: black/red, coch y bonddu: black/red/black. It was John Henderson,[5] however, who pointed out that this difference has not always been accepted and when the back history is studied, a rather fascinating story unfolds. I am again indebted to Jack Heddon for his notes on this subject.

In 1836 Ronalds[6] specifies a furnace hackle for three of his flies. For his Downhill Fly (No. 21) (called oak fly in later editions) he states, 'Legs. A furnace hackle (i.e. a red cock's hackle, with a black list up the middle, and tinged with black also at the extremities of the fibres)'. In the illustrations of all three flies, the hackle is shown as having black tips. His furnace therefore is black/red/black, but unfortunately he does not tell us what he called a hackle without black tips. His Marlow Buzz (No. 30) is given the alternative names of 'Hazel Fly, Coch-a-bonddu, Shorn Fly' and the hackle is given as 'a dark furnace cock's hackle'. The illustration of the beetle of which the Marlow Buzz is a representation is the typical black and red.

Hofland[7] in 1839 appears to be the first author to give the name of 'coch-a-bonddu' to the hackle to be used for the Coch y bonddu fly and gives the colour as 'red and black'. Francis Francis[8] gives a 'dark red hackle with a black streak up the centre' for his Coch y bonddu fly. Many other writers could be quoted.

Up to this point therefore the point accepted by all dressers was that the hackle to be used to represent the Coch y bonddu Beetle should be a dark red with a black list. Some specified that the tips should be black; others did not.

In 1885, however, G. M. Kelson,[4] who wrote extensively on salmon flies for the *Fishing Gazette*, decided that the names for these hackles should be standardised and he wrote a letter to the editor of that magazine stating that a furnace was black/red and a coch y bonddu black/red/black. He went even further, producing the names white furnace and white coch y bonddu to name the hackles that we now lump under the name of badger and even blue furnace and blue coch y bonddu. The interesting thing is that in the considerable correspondence that followed his letter, not a single person agreed with him. Fly-dressers who had been brought up in North Wales where the coch y bonddu was popular were adamant that both names could be and were used for the same hackle. One writer suggested that furnace was the English name and coch y bonddu the Welsh. Halford summed up the correspondence by writing, 'I am diffident of expressing an opinion opposite to that of so experienced a fly-dresser as Mr Kelson, but think that the majority of your readers will agree with me that the terms coch-y-bonddhu and furnace as applied to hackles are synonymous.'

When challenged Kelson admitted that he had made up the names white and blue furnace and coch y bonddu and gave as his only authority for the difference between furnace and coch y bonddu that he had samples put up by his father in 1836 which were labelled in this way. He does not examine the

THE HACKLES

THE JUNGLE FOWL HACKLES

(including a hackle of the perfect shape for a dry-fly)

KEY TO PLATE I

(from left to right)

Grey Jungle Cock — Green Jungle Cock — Ceylon Jungle Cock

Red Jungle Cock — Perfect Shape — Red Jungle Cock – eclipse

PLATE I

KEY TO PLATE II
(*from left to right*)

Ceylon Jungle Hen

Green Jungle Hen

Grey Jungle Hen

Red Jungle Hen

PLATE II

THE GUIDE TO HACKLE COLOURS

KEY TO PLATE III
(from left to right)

Pale Ginger | Medium Ginger | Dark Ginger

Pale Red | Medium Red | Dark Red

PLATE III

KEY TO PLATE IV
(*from left to right*)

Pale Yellow Badger | Medium Yellow Badger | Dark Yellow Badger

Pale Furnace | Medium Furnace | Dark Furnace

PLATE IV

KEY TO PLATE V
(*from left to right*)

Cuckoo	Creel	Ginger Grizzle
Pale Rusty Dun	Medium Rusty Dun	Dark Rusty Dun

PLATE V

KEY TO PLATE VI
(from left to right)

Pale Blue Dun — Medium Blue Dun — Dark Blue Dun

Pale Honey Dun — Medium Honey Dun — Dark Honey Dun

PLATE VI

possibility that his father like others in Wales at that time used both names for a black/red hackle with or without black tips.

The interesting point is that although no one agreed with Kelson, his names gradually came to be accepted. In 1892, when Hale[9] published his book on salmon flies, he followed Kelson's nomenclature and gave diagrams of these hackles. In 1895 Kelson's[4] own book was published and named illustrations of the hackles were again given.

Up to 1885 either name was used for a black/red or a black/red/black hackle, but from then a gradual change took place until eventually Kelson's nomenclature was used as it is today.

All this would only be of academic interest were it not for the fact that a fly-dresser when tying a fly introduced a long time ago must appreciate that this change did take place and that the black and red cock's hackle specified may not be the one which goes by that name today. The outstanding example is the famous 'Greenwell's Glory'. Courtney Williams[10] states that he had seen a letter in Canon Greenwell's own handwriting giving the tie and he gives this dressing in his book. The canon specified a coch y bonddu hackle. It is probable that a hen hackle was used for the original and that it was fished as a wet-fly. I have never seen what we now call a coch y bonddu – black/red/black – in a hen hackle and Skues stated that they did not exist. Therefore the original fly would almost certainly be black/red and a fly in the possession of the Flyfishers' Club, said to be an original, bears this out. Therefore although Canon Greenwell called it coch y bonddu, the hackle used was what we call furnace.

The furnace and coch y bonddu seem to be doomed to muddle for there is even disagreement as to how their names should be spelt. Atkinson[11] calls the game cock of this colour 'furnace' and says that it was so called because it resembled the flames rising from the black coals in a furnace. In the correspondence in the *Fishing Gazette* in 1885 it is suggested that the correct spelling is 'furness' named after the Old English Game popular in the Hundreds of Furness in Lancashire. In his book Harrison Weir[12] also supports this spelling. Both the birds and the hackles are frequently spelt furness today and although I must admit that Harrison Weir's explanation sounds to me a little more likely than Atkinson's, I have used furnace since this is the spelling that seems to have been used by all the writers in the early part of the nineteenth century.

Taverner[13] says that Ronalds[6] spelt it furniss. This, however, is not correct. In his first edition Ronalds spells it furnace, but in one of the later editions it appears as furniss due, no doubt, to an error in type-setting.

The name coch y bonddu has been written in many different ways. Every book seems to have a different spelling. I have no knowledge of the Welsh language, but Horsfall Turner[14] in an excellent article gives a very full discussion of the subject and I follow his spelling, which I am sure is correct.

We are now very used to calling a hackle with black list and red outside furnace, so I consider that we should accept it as the name for this marking and colour.

Accepting furnace for the normal bi-colour, what should we do with a furnace with black tips to the barbs? Does it deserve a special category? Should we retain the name coch y bonddu? Personally I can see no real reason for doing so. A furnace of a pale or medium red with black tips to the barbs is rare, very rare. A very dark red furnace frequently has black tips, in fact if it is a really dark furnace such as you would choose to tie a coch y bonddu fly, it will almost certainly have black tips. The badger frequently has black tips and yet no one except Kelson has thought it necessary to give this hackle a special name. Is it essential for any fly that the hackle should have black tips? Is not the shade of colour the important thing? I suggest, therefore, that we should discard the bulky name of coch y bonddu since it serves no really useful purpose and we should refer to this hackle as a 'dark furnace'. If the dresser considers that black tips are desirable, he can say so: 'dark furnace with black tips'.

Recently I have come across another version of the meaning of the word coch y bonddu. In the commercial hackle market, this can be taken to describe a cape where the tip of each hackle is black, not the tip of each barb. Since the colour of the tip of a hackle can be of no importance, since it is always tied in or cut off, this description is pointless. I suspect that it arose owing to commercial pressure to supply the demand for coch y bonddu capes. Capes where the hackles have barbs with dark tips are not common but furnace capes where the point of the hackle itself is black are not uncommon particularly in imported capes. Surely this is another reason why we should dispense with this name which has always caused confusion.

Now we come to another name which seems to have come into use in the last thirty years: 'Greenwell'. The 'Greenwell's Glory' tied by the famous Tweed fly-dresser James Wright for Canon Greenwell was, we believe, a wet-fly, but it has become so famous that the fly tied with a stiff cock hackle and fished dry is called by the same name. As I have indicated, although the original tie was written as a coch y bonddu hackle, what was certainly used was what we now call furnace. If, as we presume, the hackle was a hen hackle, then not only must it have been without black tips, but it was almost certainly a pale furnace, for while a furnace with a bright red outer is common in a cock hackle, it is very rare in a hen. The normal hen furnace hackle is brown with a black list. It is therefore almost certain that the original fly as tied by James Wright had a brown hackle with a black list, but I cannot see that there is any real reason for producing a special name for a hackle of this colour, when 'pale furnace' is equally adequate.

Commander Walker followed Kelson and lists blue furnace and blue coch y bonddu, which is surely a mistake. By blue furnace Kelson meant a hackle with a blue not a black list. This is a complete muddle. His furnace is red with black list and his white furnace white with black list, but his blue furnace changes the colour of the list not the outer colour. If a hackle with a blue list is called a blue furnace, where do the blue dun bi-colours come in? Honey dun must disappear. He says that in his blue furnace, a name which he admitted he made up, he gives the outer colour as 'sometimes grizzly, grey, a honey or a golden', in other words a honey dun. So let us finish with blue furnace. I wonder whether a honey dun hackle with blue points and which could therefore be called blue coch y bonddu has ever been seen.

Kelson also describes a 'kneecap' and Walker has followed him. This is said to be a very rare hackle with red list and edges and black between. He says that it comes from a cross between a Polish and a Malay fowl. I have never seen one and I know of no tie which specifies this hackle. If it is required, it can easily be described under off-colours, so I see no purpose in retaining this name.

Barred

Here we come to a horrible muddle of names to describe these hackles: cuckoo, grizzle, creel, cree, monkey, chinchilla, Irish grey and many others. The bars are a form of feather patterning which can appear as black or red bars, or not infrequently as a combination of both. Because it seems to be a very dominant character, it keeps appearing in a rather indefinite form in birds which have barred ancestors. This means that we can have cocks which are good examples of this class, having black and white or red and white hackles in the textbook marking or, and this is common, a cock which has hackles which show an indefinite amount of barring. In bi-colours we look for a hackle with perfect marking, a badly marked furnace or badger is of little use, but in the barred hackles a ginger which has an indefinite amount of barring is a useful and attractive hackle making a good fly. I think therefore that in our nomenclature we should allow for this difference and make a place for these indefinite but useful hackles.

9. *Cuckoo*: A white or near white hackle with black transverse bars. In a good example, the bars should be jet black or very dark blue, of equal width to the white, they should run as nearly as possible at right angles to the shaft and the black and white colours should be clean-cut, not running into each other. The typical hackle of this colour is produced by the Plymouth Rock.

   Many names have been used to describe this marking, but I think that cuckoo is the best. It is frequently used in the poultry world to describe a black and white barred bird, Cuckoo Leghorn, Cuckoo Dorking etc. and the name comes of course from the breast markings of the European Cuckoo.

   Although Herter,[15] for some reason, seemed to think that the name cuckoo was made up by Hale, this is, of course, incorrect. It is a name long used in the world of game fowl to describe a cock of this colour. Writing in 1709 Robert Howlett[16] gives a poem on the colours of the various game cocks and describes one as 'cuckoo'. Surely therefore this long-established name is the name that we should use.

The names grizzle, grizzly, Irish grey, marley, monkey etc. are frequently used for this or variations of this hackle. In the south of Scotland the name 'Grey Hen' is used for this hackle. There is a very popular series of flies tied with a cuckoo hackle and a body of various colours of silk called 'Grey Hen and Rusty', 'Grey Hen and Yellow' etc. It is possible that the 'Grey Hen' was originally a cuckoo hackle from the Scots Grey, a variety of poultry similar to the Plymouth Rock, at one time common, but now rare. I suspect, however, that the name comes from another source and that the origin of these dry-flies was a wet-fly series with the same names but tied with a blae or blue hen hackle. Just as with 'Greenwell's Glory', when dry-fly fishing became popular this successful series was tried dry and was found to catch fish. Blae cock hackles being unobtainable, a cuckoo was used as a substitute to tie the dry 'Blae hen and Yellow' and not unnaturally this in time became 'Grey Hen'. Grizzle or grizzly are well established names used in angling literature for a very long time and there are many who would argue that this is the name that should be used, and in America grizzle appears to be the normal name. On looking at the past literature, however, it seems to me that it is often used to describe the marking *on* the hackle rather than to describe the colour of the hackle itself, e.g. ginger grizzle, dun grizzle or sometimes grizzly blue dun, and I am therefore proposing that we should retain this name to describe this type of hackle.

10. *Creel*: As with cuckoo, but the cross bars are red not black. This comes from the Old English Game fowl called by this name, now a very rare variety. The shortened name cree is also used for this bird. Creel is frequently used to describe the black and white hackle, but I think that we should reserve the name creel for the red/white and call the black/white cuckoo.

    Very frequently one finds a barred cape whose hackles have all three colours, the bars being black and red, and these are often very beautiful hackles. I suggest that these should not be given a special classification, but should be called 'creel with black'.

11. *Grizzle*: I retain this name to describe a hackle of any colour which has barring distinctly or indistinctly marked. Ginger grizzle, blue dun grizzle etc.

Dun

Because of their importance in representing the upright wings of the Ephemeridae, the duns have a special place in the eye of the fly-dresser and they are the most sought after of all hackles.

A true dun hackle is a very rare one. In this country, I have only once seen a cock whose hackles could be described as pure dun as compared to blue dun, but I have recently seen hackles from America taken from cocks specially bred for hackles and I would call them pure dun not blue dun. Although this colour is mentioned in some of the old books on game cocks, it is a colour that is extremely rare in Britain and there are no examples in the collection in the Fly-fishers' Club. I do not know of any tie which specifically states that a dun hackle is required, for I think that we can safely take it that since it has always been a rare colour, where, in the old books, dun is mentioned, blue dun is intended. In a wing, however, dun is frequently given as the colour required and here dun, not blue dun, is correct. The easily obtained and perfect dun wing feather comes from the flight feather of the Starling and it is a true dun colour.

The range of duns incorporating the famous blue dun, rusty dun, honey dun etc. are the most famous of all hackles. Unfortunately blue is not a common or popular colour in poultry. It does not breed true, blue mated to blue gives a selection of blue, black and splashed chicks, splashed being basically a white or pale grey bird with a varying amount of black or dark blue markings. Blue is not a colour that is very popular on the exhibition bench and while there are a number of breeds such as Leghorn, Minorca, Wyandotte etc. which can be obtained in this colour, there is no advantage from an egg production point of view in having blue rather than black or white: so blue cocks are scarce. The blue breed which was at one time not infrequently kept for the show bench was the Andalusian and it was therefore the main source of natural blue hackles. For this reason the name 'an Andalusian hackle' was almost synonymous with 'a blue dun hackle'. The handsome Andalusian has, however, in these days of intensive egg production, become a rare breed and even this source of blue has disappeared. I do believe that it was purely the fact that it was the only breed which was in any way common, which has given the Andalusian an almost mystical quality in the eyes of fly-dressers. I may have been unlucky, but I have never seen a bird of this breed which has produced hackles of really good quality. They all tend to be very soft and this view is supported by other writers. The natural colour of the neck is very dark; not the medium to light blue which is mainly required. Both these faults could be cured by selective breeding, but it is a mistake to assume that an Andalusian hackle is a good one.

The blue in poultry is nearer a grey than a true blue and it can be matched with feathers from other birds which would never be thought of as blue. The feather from the back of a Herring Gull is almost identical in colour with the hackle from a pale blue dun cock and that from the breast of a Waterhen is very similar to a dark blue dun, but the poultry feather usually has, in addition, the brown or dun overtone.

In a blue dun, apart from the brown infusion through the hackle, there is often a yellow, gold, rusty or brassy colour on the top of the blue. This is particularly noticeable when the hackle is held above a dark background and so that the light strikes it at an angle. The colour appears as if it has been sprinkled on to the surface like gold dust. The effect is extremely beautiful and since it gives a very natural representation to a fly tied with such a hackle, these capes are much sought after.

With all these variations and shades it is not surprising that the nomenclature of the duns is even more muddled than with the other colours. It is almost impossible to find two blue dun capes which are exactly the same, and the variation in shades is endless. It is not surprising therefore that dozens of names have been produced, each attempting to describe a particular shade. The names are so numerous that it is pointless even to give examples. I believe, however, that provided we do not attempt to produce a name for every hackle, it need not be very complicated, should we agree on the main classes covered by the name dun. As with the other colours, there are two main divisions, single colours and bi-colours. Single colour covers all the hackles where the basic colour of the feather is blue dun, but this should be subdivided into the plain blue duns and those hackles where the blue dun is heavily covered with a yellow or gold shading. The bi-colour is like the furnace or badger, i.e. a hackle with a centre list and the outside a completely different colour, the division between the colours, in a good example, being clean-cut. In the case of the dun bi-colours, however, the list is blue dun not black.

If we accept these divisions, I do not think that it is very difficult to name the duns, although with so many shades and overtones it becomes a matter of opinion where one ends and the other begins.

12. *Dun*: A true dun, the colour of the primary of a Starling.

13. *Blue Dun*: A blue-grey hackle with a strong, medium or weak overtone of brown. Well likened to the colour of steel. A pale blue dun can range from practically white to a delicate blue-grey, while at the other end of the scale a very dark blue dun can be difficult to distinguish from black. The brown overtone can be practically absent in the very pale shades and also usually in the very dark specimens, but in a good example can be very strong in the medium shades. It is like the colour of steel or gun metal which has rust ingrained into it.

14. *Rusty Dun*: A blue dun hackle which looks as if it had been sprinkled with rust, i.e. reddish-gold. The colour should lie on top of the blue. They should not appear as two clean-cut separate colours as in a bi-colour although the colouring usually tends to be strongest towards the edges and the tip of the hackle. The name brassy dun is frequently used. Walker[2] gives the difference as a brassy dun looking as if it has been sprinkled with powdered brass, i.e. golden-yellow, while a rusty dun is sprinkled with rust, i.e. reddish-gold. In theory this is a nice distinction, but I have looked at a large number of these capes and even with this experience I would find it extremely difficult to decide whether a cape was rusty or brassy. As I have said, it is practically impossible to find two blue dun capes which are of exactly the same colour; the variation is endless.

    Rusty dun capes are rare. Very few fly-tyers will have seen one, so I think it is pointless to have two categories for a hackle which is so similar. I suspect that the name brassy did not arise in order to describe the hackle, but came from the cock with a 'brassy back' which would produce a hackle of this colour, and that the name rusty was used to describe a blue dun hackle sprinkled with reddish-gold; then with two names being used to describe what was really the same hackle an artificial difference was produced. I can see no point in retaining both names and as rusty dun is, I think, the name most commonly used, I retain this name only.

15. *Honey Dun*: A bi-colour where the list is blue dun and the outside yellow or honey. This is an exceedingly rare but very beautiful hackle. In a good example, the blue dun should be medium to pale in colour and the outside a bright honey, the two colours clean-cut from each other. I accept that we do have a difficulty in naming the bi-colour duns, for while in the other bi-colours the list should always be jet black, in the honey duns the list can range from very pale to very dark blue dun and the outer colour can also range from a very pale ginger to a rich copper red, sometimes called a copper dun. A hackle of this type with a medium blue list and copper outer is, as Commander Walker says, the most beautiful of all hackles, but you will be lucky if you ever see one. I consider, however, that we can get over this difficulty by making the 'pale', 'medium' or 'dark' refer to the shade of the list and the outer colour can if necessary be described in detail. For example if the hackle required for a tie is given as 'A pale honey dun with dark honey tips' we know that the hackle specified by the inventor of the fly is a honey dun where the list is pale blue dun and the outer colour is dark not pale honey. It is most unlikely that the fly-dresser who wishes to copy this fly will have such a hackle, for it is very rare, but he will know exactly what is required and will be able to substitute the nearest that he has available.

Just as 'Greenwell' is sometimes used to describe a colour, so the name 'iron blue' seems to have crept into the hackle colour nomenclature in recent years. Here again, I can see little point in using this name. We know that the delicate Iron Blue Dun, which the trout love, has very dark blue wings and it is normal to use a very dark blue dun hackle to represent it. Why therefore produce another name when dark blue dun gives the answer? I believe that this name has come into use because a natural blue dun is unobtainable and hackles are dyed to a dark inky-blue colour specially for this fly. Indeed in the list of dyes available for fly-tyers iron blue is listed as a colour, but when speaking of the natural hackle, let us remain with dark blue dun.

At one time, blue fowls must have been more common than they are today, but they have always been scarce and if you look back into the literature you will find that fly-tyers have always been searching for a source of supply of this colour. I could quote many examples, but this by Blaine[17] is typical. This was written in 1840. 'But where shall we direct him to get the invaluable dun coloured feather? Will he, for the sake of purchasing a cock or hen fowl of this shade, travel to Wales, where they are more plentiful than with us? or will he watch the markets that he may pick up the breed and cultivate it? Wherever it can be met with (for it is scarce), it should be purchased at any price, for without it there is, or rather was, scarcely a possibility of dressing the principal varieties of the duns to perfection. Now, however, where the natural feather cannot be procured by any means, we are happy to be able to add from Mr Packer, author of "The Dyer's Guide", a mode of imitating it to a considerable degree of perfection.'

It is interesting how Wales and Devon are mentioned again and again in angling literature as the best source of supply of blue cocks. In the Collection of the Flyfishers' Club in the class showing cocks' hackles there are two special sections giving specimens of Welsh and Devonshire hackles and they contain some most exquisite examples of blue and rusty blue duns.

16. *Off-colours*: When looking at hackles, I think that 90–95% can be placed in one of the fifteen categories that I have given but there are always capes whose hackles will not fall into any classification. No matter how many classes you produced, there would always be the odd men out. Some of these capes have most beautiful hackles and when one is attempting an exact reproduction they can be most useful, but

> since it is almost impossible ever to find an identical cape, such a fly cannot be copied by anyone else. I am sure therefore that it is better to keep our nomenclature as simple as possible. This will cover the majority of hackles and will allow the inventor to indicate easily the type of hackle to be used. If a hackle will not fall into any of these classes, it can be treated as an off-colour and described.

In all hackles it is very desirable to have the colour of the back as near that of the front as possible. The reverse side unfortunately often has a chalky colour. This stems from the Red Jungle Cock, the ancestor of our domestic fowl, which has a very chalky reverse. It is a characteristic that could be cured by selective breeding. Many years ago there was a breeder who disliked this type of hackle and settled down to produce a strain which had hackles of the same colour on both sides and it is said that he eventually succeeded.

In the following table, I have summarised the sixteen classes that I have suggested that we should use to describe our hackles so that others may know which colour and marking is intended for a certain fly. It is obvious that we can and will subdivide almost all these classes by adding 'pale', 'medium' or 'dark', indeed we can go further if we wish to do so by adding 'very pale' and 'very dark', but I do not think that for practical purposes this is necessary.

Hackle Classification

No.	*Name*	*Description*
Single colours		
1	White	Pure white.
2	Cream	From off-white to dark cream.
3	Ginger	From dark cream to pale red.
4	Red	From pale red to dark red.
5	Black	Jet black.
Bi-colours		
6	Badger	Black list, outer white or cream. Tips of barbs sometimes black.
7	Yellow badger	Black list, outer from cream to pale or medium ginger. Tips of barbs sometimes black.

Bi-colours (contd.)		
8	Furnace	Black list, outer from ginger to dark red. Tips of barbs sometimes black.
Barred		
9	Cuckoo	Black and white bars. Pale to dark can be used to describe the general coloration of the whole hackle.
10	Creel	As cuckoo but ginger or red replaces the black.
11	Grizzle	Not a colour on its own, but a description of the marking, e.g. ginger grizzle is ginger with a certain amount of barring showing.
Duns		
12	Dun	A pale dun colour all over.
13	Blue Dun	From pale grey to near black. Brownish overtone to a lesser or greater amount.
14	Rusty Dun	A blue dun spangled with rust colour. If pale, medium or dark is used this should refer to the shade of the blue dun.
15	Honey Dun	Blue dun list, outer normally honey colour but can be from cream to near red. If pale, medium or dark is used, this should refer to the shade of the blue list, the shade of the outer colour to be described if required.
Off-colours		
16		All other hackles which cannot be placed in any of the classes and have to be described.

I have said nothing about dyed hackles, for this surely is only tolerated if the fly-dresser is unable to obtain the natural colour. The first step in dyeing a hackle is to steep it in a strong soap solution and then in near boiling water to remove the natural oil, otherwise the dye will not stain properly. This may be admissible and indeed an advantage for a wet-fly where quick penetration through the surface film is an advantage, but for a dry-fly, once the natural oil is removed you have a poor soft object. There is, however, one colour which the natural hackle cannot represent and that is the 'olives'. No hackle has this shade. White and cream are usually used and dyed an olive colour and this is the best that can normally be done, but there is one way of producing a perfect 'olive' very, very near to the natural colour. Unfortunately to obtain this perfect representation, you have to start with that rare hackle a natural blue dun and the shade is obtained by staining it in picric acid. Picric acid was at one time used as a treatment for burns and could be purchased from any chemist, but is now not easily obtained. In dry crystalline form, it is, I understand, explosive and used in the explosives industry, but when dissolved in water it is quite harmless. Wet your blue dun hackle in cold water and then soak it in a solution of cold picric acid. Take it out and wash it under the cold tap every now and again to see how you are getting on, returning it to the solution if necessary. The yellow picric stain 'takes' on the natural hackle; you do not remove the natural oil and you end up with a most beautiful olive.

Recently an interesting method has been invented of artificially producing blue dun hackles. The treatment, which is fully described by Leiser,[18] consists of dipping the cape in silver nitrate and then 'developing' it as a photographic negative. This method is still new and time will tell whether it proves to be successful. On the good side the natural oil in the hackle is not removed, but for useful blue dun hackles for dry-flies, white capes of high quality would be required and unfortunately in white the quality always seems to be only moderate. This difficulty does not, however, apply to hen capes for wet-flies.

I should like to end this chapter by quoting from Cutcliffe[19]: 'Of the different varieties of feathers as I have said, the rusty are the most valuable in all respects, and therefore we must seek to procure a good assortment of rusty blues; a series of shades in colour resembling exactly the hare's back, these feathers are beautifully spangled in yellow and gold, on a dark blue grounding, they are very scarce, and of all feathers the best for rapid stream fishing; also rusty reds – not forgetting the blood red, the black red, or a feather whose central rib, apex, and base is black, whose back or hinder surface is shining and dark, and all the intervening part dark blood-red colour; this of red feathers next to rusty is the best: then come the browns, serviceable in the early months: and the blues, from the pale, sober blue of the heron's wing, to the dark rich spangled blue of the back of a water rat, as seen when the entire fur is smoothed down. A few black, and a few white, and our assortment is complete.'

References

1. Cross, Reuben R., *The Complete Fly Tier*, U.S.A., 1936. Reprinted 1971.
2. Walker, C. F., *Fly-tying as an Art*, London, 1957.
3. Elder, H. F. D., 'The Fly Fishers' Club Collection of Fly Tying Material', *Flyfishers' Journal*, Vol. 64, Autumn 1975.
4. Kelson, George M., *The Salmon Fly*, London, 1895. *The Fishing Gazette*, 1885.
5. Henderson, J., 'Cochybonddu and Furnace Hackles', *Flyfishers' Journal*, Vol. 25, Spring 1936.
6. Ronalds, Alfred, *The Fly-Fisher's Entomology*, London, 1836.
7. Hofland, T. C., *The British Angler's Manual* . . ., London, 1839.
8. Francis, F., *A Book on Angling* . . ., 1867; 5th edition, London, 1880.
9. Hale, J. H., *How to Tie Salmon Flies*, London, 1892.
10. Williams, A. Courtney, *A Dictionary of Trout Flies*, 2nd edition, London, 1950.
11. Atkinson, Herbert, *The Old English Game Fowl*, London, 1891.
12. Weir, Harrison, *Our Poultry*, 1902.
13. Taverner, Eric, *Fly-tying for Trout*, London, N.D. [1942].
14. Turner, Eric Horsfall, 'Cocky Bonne Dee', *Flyfishers' Journal*, Vol. 58, Winter 1969.
15. Herter, G. L., *Professional Fly Tying* . . ., 19th edition, U.S.A., 1971.
16. Howlett, R. H., *The Royal Pastime of Cockfighting*, London, 1709.
17. Blaine, Delabere P., *An Encyclopaedia of Rural Sports*, London, 1840.
18. Leiser, Eric, *Fly-Tying Materials*, U.S.A., 1973.
19. Cutcliffe, H. C., *The Art of Trout Fishing on Rapid Streams*, South Molton, 1863.

CHAPTER VI

Quality

Almost every book which deals with dry-fly fishing has something to say about the necessity for a really good hackle to produce a good fly, but apart from saying that the hackle should be stiff and should sparkle in the light when twisted in the fingers, there is no clear definition of what is required. Another test often given is that if the tip of a good hackle is bent, with the butt firmly held in the fingers, when released it should spring back to the original position as with a good dry-fly rod.

What are the essential differences which separate a good from a moderate dry-fly hackle? Commander Walker,[1] although he gave a list of points which are required, rightly said that it is easier to recognise than to define a good hackle, but even here it is a question of standards. How can you judge if you have never seen a first-class hackle? Many experienced fly-tyers are in this position, for really fine hackles are very, very rare. A further difficulty is that what one tyer calls good, another may discard as useless if he has been fortunate enough to be able to use the very best. The search for hackles of first-class quality is not new, and even in 1847 Stoddart[2] said that not one cock in ten walking the farmyard produced hackles of good quality.

In the next chapter I have tried to differentiate between good and bad shape in a hackle; this is much simpler since actual measurements can be used to classify the standard for shape. In many books good shape has been taken as being an essential part of good quality, but I prefer to split these into two. By 'quality' I mean the physical features of the hackle, the stiffness, the strength of the individual barbs, the polish, the sparkle, and I regard 'shape' as something separate. The two do not necessarily go together, for hackles can have good quality and bad shape and vice versa. In my strain of cocks, I believe that the shape that I have now produced is exactly what I require, but I am well aware that I have not reached perfection in quality.

Regrettably, I have been unable to improve upon the vague definitions of 'quality' contained in the fishing books. It is a matter of comparison, but if we consider the functions of the hackle in a dry-fly and the reasons why a 'stiff' hackle is superior to a 'soft' one, then we will have a good understanding as to what is wanted and of the standards that we should look for.

The hackle on a dry-fly must fulfil three functions. First, it must support the fly on top of the surface film. Second, it must produce the correct light pattern as seen by the trout on the under surface of the water. Third, if the fly is a 'hackled', as compared with a 'winged', fly it must produce an illusion of wings which is good enough to satisfy the trout if he is inspecting the fly through his 'window'.

To deal with these in the reverse order. The illusion of wings does not really come into our terms of reference. To get a good answer on this, someone would have to view from underwater a series of artificial and natural flies, the artificial tied with various materials to represent wings, and then to decide which was the most accurate imitation. Having produced an answer, it would have to be tried out to see whether the trout agreed. Until this has been properly done, all I can say is that viewed with the human eye from above water, an artificial hackled fly tied with a hackle of a colour very like the live insect, is extraordinarily like the real thing. At short range the upright barbs do look very like wings and it is often extremely difficult during a large hatch to pick out your own artificial fly. The trout, however, gives the real answer and a 'hackled' fly, because of its results, is the most popular type of fly with the present-day dry-fly angler. The solid-wing fly is not as often used now as it was in the past.

The second purpose of a hackle is to produce the correct light pattern as seen by the trout on the under surface of the water. Again what we require is a detailed series of experiments viewing flies, artificial and natural, from underneath. It is a good number of years since Harding[3] and Lane[4] carried out their interesting experiments. Cameras have improved tremendously. Accurate colour photography is now possible and it should not be too difficult to produce a series of films in colour which would show the trout's view as a natural fly is carried towards him by the current and then to compare this with a range of artificial flies tied with various materials. It should be fairly simple to decide what is required for a good representation as seen by the human eye. To find out how anything appears to a trout is much more difficult, but we could at least start with a good representation as we see it and then obtain the reaction of the trout. Until these experiments are again carried out, all we can do is accept the findings of Harding,[3] Lane,[4] Ward[5] and Hewitt.[6] They all appear to be adamant that the light pattern on the under surface produced by the feet of a live dun is best represented by the points of a really stiff cock's hackle. As Hewitt[6] has pointed out, the feet of the dun draw up the water by surface tension to create bright blobs or lines of intensity which must be clearly visible to the trout long before the fly reaches his position. They all stress the importance of producing the correct light pattern and all agree that while a really stiff hackle produces a light pattern which is very like that produced by a live dun, a soft hackle, owing to the flimsiness of its barbs, allows the fly to sit much lower, resulting in a blotchy and unnatural pattern. A stiff hackle certainly appears to produce a satisfactory answer for if a trout is feeding steadily on newly hatched duns as they sail down the stream, a fly tied with a good hackle, if properly presented, is usually accepted.

This brings us to the first and perhaps the most important reason for a good hackle. It must support the fly, on top of, not floating in, the surface film and it must be remembered that the fly contains a relatively heavy steel hook. Again I stress that a first-class representation must sit on top of the water like the live dun.

The nymph breaks through the surface film, its skin splits and out comes the dun. It sits on the top of the surface film supported on its six feet. It is amazing how large numbers sail unharmed down the current waiting for their wings to dry when they can then fly to the bank. The sudden swirl or broken water, however, ends the life of many of the duns, for if they break through the surface film and become wet they are trapped and can never get back onto the surface alive. During a large hatch of duns, if you look in the backwaters, you will see hundreds of bedraggled insects which are struggling in the surface film or are floating dead. The evidence we see in these backwaters adds extra weight to the thought that the live dun sitting on the surface is much more attractive to the trout than one struggling or floating in the surface film. In these backwaters there is plenty of food for a trout, but the fish never seem to find these trapped insects very attractive. They will lie at the edge of the backwater, where the current brings along a steady stream of duns, indeed this is a favourite position as every fisher knows, but they rarely cruise through the backwaters where there would appear to be an abundance of tasty morsels. With spinners, however, the dying 'spent' spinner lying in the surface film attracts the trout, possibly because, since the live spinner rarely sits on the surface, this is normally the only time it can be taken by the trout. Indeed I find it strange that there are so many dressings, which are supposed to represent spinners, where the hackle specified is a stiff cock hackle and therefore intended to keep the fly sitting on the top of the water. As far as I have been able to ascertain the exact method of egg-laying by every species does not seem to have been worked out, but according to Harris[7] and Goddard[8] there are two ways, which are adopted by the majority of the up-winged flies. After mating, except for the males of the Mayfly or Greendrake and to a lesser extent the Iron Blue, the male spinners rarely return to the water and there is therefore really no purpose in tying a fly to represent them. The male spinner of the Mayfly frequently falls spent on the surface, but I have not enough knowledge of this species to say whether it is commonly seen on the surface in the up-wing position. The male spinners of the Iron Blue are supposed to return to the water, sometimes in large numbers; this I have not seen, but I understand that if it does happen they appear in the spent position in the surface film with wings outstretched.

The female spinners must return to the water to deposit their eggs and they do this either by flying low over the water and dropping their eggs, which sink to the bottom, or they land on a solid object and creep down into the water. Their wings are pressed back against their body and they are encased in a bubble of air, which they take down with them where they attach their eggs to a solid body underwater. Whichever method is used, after egg-laying the life of the female spinner is over and she quickly dies.

With the species that lay their eggs when above the surface some may drop them from above the water, but others may touch the surface at the moment of egg-laying. After egg-laying the majority fall spent with wings outstretched on to the surface. Dr Orr of Canterbury informs me, however, that he has frequently seen female spinners of the Blue Winged Olive coming down the fast water, where their eggs are dropped, in the up-winged position, but that they quickly collapse with wings outstretched. He has also recorded Medium Olive spinners and Pale Watery spinners in the up-winged position.

With the species that creep under the water to lay their eggs, even if a few do regain the surface in their bubble of air, very few if any can ever be seen sitting on the surface like a dun.

It would therefore appear that the trout normally sees the spinner dying in the spent position with wings outstretched lying in, not sitting on, the surface film. Why then does one ever tie a representation of a spinner with a stiff cock's hackle? Surely the correct representation should be tied with spent wings and no hackle at all. The famous Pheasant Tail spinner may take a lot of trout, but I cannot believe that they take it as a spinner. I am sure that Harris[7] must be correct when he advocates that flies should be of two distinct types, either sitting above the surface on stiff hackles to represent the insect as a dun or lying flat in or on the surface with little or no hackle to represent the dying spinner.

The good dry-fly hackle must have barbs which are stiff enough to support the fly right on the top of the surface, cast after cast. For this it is essential that a hackle must have barbs of the correct shape and length. There is a type of hackle where the barb is relatively fine at the base, tapering slowly to the tip. This does not produce a good fly, since these thin barbs tend to be soft and they bend. The perfect barb is relatively broad at the base and short in length; this gives a stiff barb which is capable of supporting the hook. Owing to the difficulty of obtaining really stiff hackles for dry-flies, people sometimes resort to using a hackle with longer barbs and then clipping them to a shorter length. I believe that there is even an apparatus on the market for doing this. The result is a fly where the barbs do not end in a point and while I am sure that it is more satisfactory than a soft hackle, it is a clumsy thing to look at and can never give the satisfaction of a fly dressed with a good stiff hackle.

Once a natural fly gets wet it is finished, but our artificial has to stand up to very rough treatment and still be dry when it lands on the water at the next cast. Angling books speak about the fly being picked off the water at the end of each cast and talk about 'should the fly be drawn under the water in a rough stream . . .'. I wonder whether any of these writers has ever carefully observed what happens to their fly in the normal actions of dry-fly fishing, when the fly is cast, allowed to float a certain distance and cast again. If he watches carefully he will see that the fly is drawn under the water or through the surface film at the commencement of each cast. It is impossible to pick it cleanly off the surface. To allow this to happen none of the line or cast would have to touch the water. It is possible in dapping where the light floss line is held above the surface by the wind and a lift of the rod will lift the fly off the surface, but where the line and cast lie on the surface a clean lift is impossible. The action of shortening the line and the accelerating movement of the fly, which is the first part of the back cast, means that the fly is drawn through or under the surface. The upholders of casts made of materials which sink, natural gut, for example, are also making quite sure that their fly gets a good soaking every cast. Even the modern floating lines claim that the last few inches or feet of the tip sink and that this makes the line less visible to the trout. This ensures the same result. Even if the whole line and cast were devised to lie on the surface and never break through the surface film, the shortening of the line, the first part of casting, would draw the fly across the surface and its irregular outline would inevitably make it break through the film and get wet. The cast and line are not, however, lying on the surface; a certain length is in or under the surface and the first shortening draws the fly along the same path and therefore under the water. Indeed, if when casting you listen carefully you will often hear the sucking noise made by the fly as it is plucked up through the water. The longer the length of line cast and the greater the distance the fly is allowed to float, thus giving the cast and tip more time to sink, the more certain you are of drawing your fly well and truly under the surface.

If you purchase a dry-fly, why is it that it is always tied on a hook with an upturned eye? Even the catalogues offering materials for fly-tyers list hooks for dry-flies with upturned eyes and those for wet-flies with downturned. Do we believe that an up-eyed hook allows one to pick the fly off the surface, while a down-eyed helps to drag the fly even deeper?

The position therefore is not, "*if* the fly gets wet . . .", but "*since* the fly is immersed at the end of each cast. . . ." Yet the fly must be dry once again at the beginning of the next cast.

Why is it essential that it should be dry? A scientific explanation as to what happens when an object sits on top of the surface and why it cannot remain in this position if it is wet, is not required, but a simple explanation is necessary to understand the mechanics. The well known party trick of the needle and the glass of water is such a good example of the principle involved that I will describe it. Take a glass of water, an ordinary sewing needle and a piece of blotting or absorbent paper slightly larger than the needle. Place the paper, with the needle on it, flat on the surface of the water. The paper will lie on the surface supporting the needle, but gradually water will be absorbed by the paper, the paper will sink, but the needle will be left lying on the surface of the water. A heavy piece of metal floating on water! Gently touch one end of the needle so that it is forced through the surface film and the whole needle will slip through the surface and sink. Retrieve the paper from the bottom of the glass and try to place it on the surface film: it will immediately sink, but a new piece which is dry will sit on the surface until it gets wet.

This simple children's party trick illustrates the principle of surface tension. Any liquid behaves as if it has an invisible skin on the surface which prevents a dry object from penetrating. This 'skin', however, is not very strong and an object of any weight will break through and sink. A dry object can be supported on top of the surface film but if it is wet, the liquid flows over the material and it sinks.

The live dun sits on the surface film. The only points which are in contact with the water are the tips of its six legs and these are not in the water but on top of the surface film. Our dry-fly, if it is to be a true imitation of the live dun, must behave in the same manner. It must sit on the surface film and to do this it must be dry. The artificial fly having landed on the surface and having been allowed to travel a certain distance down the current is then pulled under as it is lifted for the next cast. It is bound to get wet, and yet it must be dry by the time it lands once again on the surface. Casting as short a line as possible and also allowing the fly to travel only a short distance before the next cast will help to prevent the fly from becoming well soaked each time, but no matter what we do, the fly will have a certain amount of water on it as it leaves the surface. This must be removed, but how can this be done? The angler achieves this or attempts to achieve this in three ways. False casting, by applying a 'floatant' (as it is normally called), and by using really stiff hackles.

False casting. There are two actions here. First the physical action of throwing the particles of water off the surface of the fly by centrifugal force and second the actual drying or removal of the water as vapour as happens in the drying of clothes in a strong wind. The first is the most important action, but in order to achieve the best result, the particles of water must be actually lying on the surface of the hackles from where they can be thrown off by force. Once the hackle or body material has become really wet and has absorbed water, it is only the surplus that can be thrown off in this way and to become dry the water must leave it as vapour. While the passage through the air will certainly assist this drying process, once the fly does become wet, false casting will not fully dry it out and before long it will have to be retrieved and dried out in the hand.

The second thing we do in our attempt to keep the fly dry is to treat it with some chemical which will prevent it from absorbing water. This is usually called a dry-fly dressing or floatant. It is very unfortunate that the name 'floatant' was ever used for this type of liquid dressing, for it has given a completely wrong idea as to its purpose. People naturally think that these dressings are applied to help the fly to *float* and the fact that one of the first dressings to be used was paraffin oil has added to the misconception; oil floats on water so the oil was put on the fly to make it *float*. This has even helped to establish the idea that the fly should float on the surface, when it should of course *sit* on the surface. The truth is that the fly-dressings certainly do help to keep a fly dry, but not by making it float; their action is that they are water repellants and a fly coated with such a chemical does not absorb water, the water running off the surface and any remaining on the surface of the fur or feather does so as droplets which can be thrown off during the false cast. We have come quite a long way from the original paraffin oil dressings; the chemists have produced some very effective chemicals, but it is a difficult problem with which they are faced. A dry-fly is a very delicate thing with filaments of feather and fur and the dressing must not clog these fine fibres. The dressing must be thin, easy to apply and if its action depends on drying before use, it must dry very quickly. It must remain coating every fibre when whisked at high speed through the air and, most difficult of all, it must withstand being pulled through the water at every cast. Efficient chemicals have been produced and they will get better, but we have a long way to go before we can hope to have a dressing which is not only completely water repellant but will be so permanent that it will prevent the fly from absorbing water throughout its whole life. The present-day dressings do get washed off the fly and the fly must be re-treated at intervals. Needless to say, it is essential that the fly should be completely dried out before re-treating. A rough squeeze in a handkerchief, a heavy dose of fly-dressing and start fishing again is not really very much use. The old idea that the dressing helps the fly to *float* once again. We all know how difficult it is to get a fly really dry. No matter how carefully we dry it, it never sits up like a fly fresh from the box. I am sure that those who frequently tie on a fresh fly are wise. Halford[9] said that the first cast with a new fly was the one most likely to catch a trout and that the greatest care should therefore be taken with this cast.

The third and most important way in which a fly can be kept dry for as long as possible is to use a hackle with really stiff barbs. We have seen that stiff barbs give the correct surface pattern and also allow the fly to sit right on the surface, but to remain effective in this way, the barbs must remain dry. Any material once it is wet tends to become limp and has not the same strength and stiffness as when dry. A piece of paper is a good example. How can we keep our hackle dry when it is drawn through or under the surface with every cast? It is of course an impossible task, but if we use a really good stiff hackle compared with a soft one it will stand up to this treatment for quite a long time. If the barbs are stiff, the fly will be kept above the surface of the water and it is only at the moment of casting that the hackle can get wet. If the hackle is soft, the fly sags into the surface film and both the barbs and the body can absorb water all the time. Another reason is that every hackle has a certain number of barbules attached to the barbs. In a hen's hackle there are barbules right along the length of each barb. In a poor cock's hackle the barbules extend well out from the shaft giving the hackle quite a wide 'web'. In a first-class hackle this web is very narrow, the barbules extending along the barbs for only a short distance. The barbules, which are tiny hooks, hold the barbs together to make a solid feather and, in the body feathers, they also trap the air and this helps to keep the bird warm. These barbules, however, are exactly what is not wanted in a dry-fly, since they trap and hold minute droplets of water preventing them from being thrown off during the false cast. A good dry-fly hackle therefore is one with as few barbules as possible and this is exactly what a fine cock hackle will give. A final point. The hackles in a cock have been evolved as display feathers, they have become bright and shiny so as to be shown off to the best advantage. A shiny feather must not only have few barbules, for they would give it a dull appearance, but each individual barb must have a bright and shiny surface. This is exactly what is wanted for a dry-fly. The hard shiny surface makes the barb not only stiff to support the fly, but the polished surface does not absorb water, which is then easily thrown off. For all the reasons given, nothing has been found as effective as a really fine stiff cock's hackle for supporting a dry-fly. No one has produced an artificial product which will be an alternative and it is certainly difficult to imagine that anything fine and strong enough could be produced.

Oliver Kite[10] gives a further reason for the superiority of a stiff compared to a soft hackle. He considers that a stiff hackle holds the fly above the surface supported on its fine points and that the airs and breeze rock it, simulating life.

Many writers have pointed out that there is a definite relationship between stiffness and colour. Black hackles always tend to be soft. The reds are good and for some reason badgers are also good, which is strange for whites are usually poor. Blue duns, being near black, are also very difficult to obtain in satisfactory quality. It has often been pointed out that duns are usually very disappointing, being weak and long in the barb. Commander Walker[1] goes the length of saying that in thirty years of fly-tying he has never seen a black hackle fit for a floating fly. I think that some of my birds can do better than this, but to date I have not bred a black which I would class as top quality.

The number of barbs on a hackle is a matter of some interest. As one would expect, a small hackle tends to have a greater number of barbs per centimetre of length than a large one, but what is possibly surprising, in hackles of the same length there can be a large variation in the number of barbs. In a hackle of about 4 cm of total length, there can be 19 barbs per centimetre or as many as 28. The easiest method that I have found of counting the barbs is to draw a straight line on a piece of card and on this line mark off a centimetre. Lay the hackle with its shaft on this line and ensure that the measured centimetre is approximately in the centre of the useful portion of the hackle. Holding the hackle firmly in position and using a magnifying glass, it is quite easy to count the number of barbs between the two marks. It is not necessary to spread out the barbs if a good glass is used, and the point of a sharp needle helps one to count. The figure that I give is the number of barbs per centimetre on one side of the shaft.

I have counted a great number of hackles and have attempted to analyse the results to see whether the number of barbs can in any way be used as a guide to quality, but I have been unable to obtain any correlation with either quality or colour.

It would be expected that when comparing a long with a short hackle, since the barbs on the first are longer than those on the second, they must be broader at the base, must take up more room on the shaft and therefore there must be fewer in a given length. This is true. Comparing the number of barbs per centimetre from a series of hackles taken from the top to bottom, down the centre line of a cape, there is a reduction in the number. But there does not appear to be a constant relationship, the number of barbs increasing fairly rapidly as the hackles get smaller at the top of the cape. Between hackles of the same size taken from different capes, however, there is the variation in number that I have mentioned. Walker[1] pointed out this difference in the number of barbs and stated that this certainly, in some cases, depended on the colour of the hackle. He said that a white hackle had up to 10 more barbs per inch (2·5 cm) than a black of the same size and 8 more than a blood red or a dark blue. I have looked carefully at my data to see whether I can establish this point, but cannot find it to be true. While the number of barbs varies from cape to cape, this variation does not appear to depend in any way on the colour. I suspect that the variety of the poultry is important, because for example all the hackles from the wild Red Jungle Fowl that I have counted have the number of barbs per centimetre well above average.

Both Walker[1] and Reuben Cross[11] suggest that a hackle of good quality has a greater number of barbs than a poor one. Cross compares a bad, when spread out from the centre web, as being like the teeth of a coarse comb, while there should be a great number of teeth as in a fine comb. I have looked at my hackles from this point of view and can again find no counts to support this suggestion, I have hackles which I would call of very good quality being the 20 barbs/cm class in a hackle of 4 cm. The possible exception again being in the wild Jungle Cock where hackles of very high quality always have a high number of barbs.

I cannot therefore find any evidence that the number of barbs in any way depends on the colour or that a high number of barbs denotes high quality. Indeed I would suggest that in theory the ideal hackle should have short barbs but that they should be broad at the base and taper quickly to the point. This must give a very stiff barb, but because of the relatively broad base to each barb, there cannot be a very high number of barbs in a given length.

We now come to the interesting and often repeated statement that to produce good hackles a cock must be two or three years old. I have traced this back in angling books for more than a hundred years. Sometimes two years is given as the minimum age, sometimes three or more. When I started to breed game cocks I found that the hackles from my young cocks were most disappointing, being soft with long barbs. However, all the books said that the Old English Game had perfect dry-fly hackles, so it was obvious that these hackles were simply the normal soft hackles produced by a young cock and that when it was older I would have a supply of really fine hackles. When after the second and third moult I could see no difference, I began to wonder whether age had indeed any effect on quality. As each successive year's hatching produced the same result, I was puzzled. That the quality of hackle varies from one cock to another is quite certain. The hackles from two cocks produced from the same clutch of eggs can vary tremendously. The one can have very stiff and the other very soft hackles, but I have found that with my birds this quality does not change no matter for how many years the birds are kept. The bird with stiff hackles will always have stiff hackles and they do not improve as the bird gets older. The bird with soft hackles will have soft hackles all its life. I have found that the stiffness of hackle is a hereditary factor and that with selective breeding for hackle quality the stiffness can be improved. It is a hereditary, not an age factor. However, this statement has been repeated so often and has become so well known, that it cannot be lightly dismissed as being an 'old wives' tale' without careful examination. I must say that in a more detailed study of the literature, I have not found it repeated by anyone, except Skues, who I know had themselves bred cocks for hackles. Indeed at the end of Hughes'[12] article on breeding poultry for hackles he says that a young cock can produce first-class hackles. He words it almost as if he also has found no truth in this statement, but does not wish to go the length of saying so. Hughes knew as much or more about breeding for hackles as anyone before or since. One of the difficulties is that the only people who can say true or false are people who have bred poultry themselves or have a range of hackles taken from cocks of known age, and there are not many who have this experience.

John Henderson started by believing that the quality of the hackles that he bred improved as the bird grew older, influenced no doubt by what had been repeated so often that it had become accepted fact. After further experience he came to the conclusion that with his birds it was not true.

On the other side there are numerous writers who have stated that the quality does change with age. The first writer that I have found that discusses the subject in detail is Pulman.[13] He obviously kept or knew a lot about keeping poultry. In his *Vade-Mecum* of 1846 he mentions the change in quality as the bird grows older and he enlarges on this in his 1851 edition. However, although he considers that the hackles of both cock and hen change as the bird grows older, he prefers the hackles from a cock 'between ten and twenty months' and says that 'The common faults are, a stiffness and coarseness of the stem and of the plane fibres – invariably the case with the hackles of an old cock'; on the other hand he says, 'As to hens' hackles, it must be remembered that they should never be taken from a hen less than two years old – an age when the cock, as a hackle producer, is becoming worthless. Hackles from a young hen are always brittle in the stem, particularly at the point, and the plane fibres are of too soft and downy a nature'; and later, 'Hens, unlike cocks, improve for the fly-maker with age.' Pulman used cock's hackles for his hackle flies 'such as palmers', and hen hackles for 'wing-flies'. He considered the hackle one of the most important parts of the fly and one could not take too much trouble in choosing a good hackle. He speaks at some length on the necessity of a fine glossy hackle. It cannot be taken for granted that he was talking about 'wet' flies, for the 'dry' fly was not unknown to Pulman. The interesting fact is that Pulman, who appears to know a lot about hackles, went out of his way to discuss the effect of age on hackle quality both in cocks and hens, although at the end of it he advises the reverse of what everyone else has said since. The important point, however, is that he did consider that quality changed with age.

Halford[9] advises hackles from a cock of 'eighteen months to two years' although he does not say that those from a younger bird are of poorer quality. I do not think that he himself kept poultry for hackles, but he says that he had 'succeeded in persuading a relative to perpetuate a good strain of Blue Andalusians'.

Cutcliffe,[14] who loved good hackles, kept fowls and although he discusses colour, quality, breeding and management at some length, he never suggests that age has any effect on quality.

Skues[15] wrote that a cock of two or more years produced a better hackle than a young bird and although he certainly did at one time keep or had birds kept for him, I do not think that, even with all his knowledge of hackles, he was ever a serious breeder. I have only found one reference to his keeping poultry for hackles, this being in a letter to Major Evans. If he had been a serious breeder, I am sure that this would have been mentioned frequently in his articles and books. It is probable therefore that he also was simply repeating what had become to be an accepted fact.

Since I could not find this statement true with my own birds and therefore began to doubt whether it really was true, I wrote, in 1974, to the angling journals asking whether anyone could produce evidence for or against this oft-repeated saying. I did not receive many replies to my letter, which was to be expected because, as I have said, only someone who can compare hackles taken from birds of known age can give an opinion which is of any real value. You can compare hundreds of capes, but unless you know the age of the bird when it was killed, you are no further forward. The one or two who replied and who had kept cocks for hackles all supported my suggestion that there was no evidence to indicate that the quality does change with age. Colour, yes. From year to year, the colour of a cock's hackle can change, usually they tend to get paler although possibly examples of the opposite can be found, but not quality. One or two disagreed with me considering that age had an effect on quality, but when their case was examined they really were repeating the old story and had no evidence to support their contention. As an example, one fly-tyer sent a most beautiful hackle taken from a four-year-old cock, a hackle of first-class quality and he considered that this proved the case. Unfortunately, however, he did not have any hackles taken from this cock in its earlier years so this proved nothing, for the hackles in its first year might have been equally good.

This statement has been repeated so often that it is now accepted as fact. It would therefore take a lot more evidence than any single breeder can produce before we can say that it is not true. All I can say is that with my strain of birds, I can see no difference in the quality of hackle produced by a cock in its first year from that produced by the same bird in later years and that I can find no detailed information in angling literature which makes me believe that my findings are unusual.

There is one fact that makes me particularly cautious in making any statement regarding age and quality. In his *Pheasants of the World*, Delacour[16] states that the hackles produced by the Red Jungle Fowl in its first winter plumage, which is really its first full plumage, are less bright and shorter than the adult. If there is a difference in the first year of the Jungle Fowl, the ancestor of our domestic bird, is it not possible that some strains carry this on?

If I am correct and there is indeed normally no difference, how or why did this statement arise? The young cock after the chick down disappears grows two sets of hackles before the full adult hackles are produced. As the second set grows, the stage has been reached where the cocks and the hens can be distinguished by plumage, the shapes of the tail feathers change, the saddle hackles begin to appear and the neck hackles on the young cock become pointed. It would be at this stage in earlier years that the young cocks on the farm would be earmarked for killing off as soon as they were of a reasonable size. They would not, however, be killed off immediately for, having fed them to this stage, it was worth the cost of a little more food to obtain a cockerel of reasonable size for the table. The hackles at this stage although pointed are very soft and of no use for a dry-fly; it is the adult hackles which only start growth at this time that are required. Is it possible that experienced fly-tyers warned others against the quality of the hackles obtained from these young immature cockerels which were on sale in the markets and that this warning has become wrongly repeated as 'a hackle from a first-year cock is of poor quality' which in turn easily becomes 'a good hackle only comes from an old bird'. This is only a possible explanation as to how this saying came about, but I believe that it may be the correct one.

Recently some articles have appeared in the angling journals suggesting that for dry-fly fishing a soft hackle is better than a stiff one. The authors produce some ingenious arguments to support their case and I suspect that some of these articles are written simply to produce controversy. Harding[3] states that looking from underwater a soft hackle gives a very unlifelike shape. Lane[4] says the same, but the most important thing is that almost every writer since the days of Halford has indicated that the essential part of a good dry-fly is a stiff hackle. They found this out not in theory but in actual fishing on the river and base it on the reactions of the trout. I must say that I agree. For the pleasure obtained in fishing and in the actual results achieved, there is all the difference between fishing with a stiff and a soft hackle. There is great pleasure in fishing a fly with a good hackle and little pleasure in fishing a fly with a poor one. Woolley and Williams do, I think, sum it up. Woolley[17] says, '. . . there is no doubt that a good stiff game hackle enhances sport', and Williams,[18] '. . . the difference between a good hackle and a poor one often represents the difference between a full creel and an empty one'.

References

1. Walker, C. F., *Fly-tying as an Art*, London, 1957.
2. Stoddart, T. T., *The Angler's Companion . . .* , Edinburgh, 1847.
3. Harding, Col. E. W., *The Flyfisher and the Trout's Point of View*, London, 1931.
4. Lane, J., *Lake and Loch Fishing for Trout*, London, 1955.
5. Ward, Dr Francis, *Marvels of Fish Life*, London, 1912.
6. Hewitt, E. R., *Secrets of the Salmon*, New York and London, 1925.
7. Harris, J. R., *An Angler's Entomology*, 2nd edition, London, 1956.
8. Goddard, J., *Trout Fly Recognition*, London, 1966.
9. Halford, F. M., *Floating Flies and How to Dress Them*, New York, 1886.
10. Kite, Oliver, *A Fisherman's Diary*, 1969.
11. Cross, Reuben R., *The Complete Fly Tier*, U.S.A., reprinted 1971.
12. Hughes, Thomas, 'How to Breed Fowls for Fly Tying', *Fishing Gazette*, 1922.
13. Pulman, G. P. R., *The Vade Mecum of Fly Fishing for Trout*, 1st edition, London, 1846; 3rd edition, London, 1851.

14. Cutcliffe, H. C., *The Art of Trout Fishing on Rapid Streams*, South Molton, 1863.
15. 'V.C.' [Skues, G. E. M.], *Silk, Fur and Feather . . .*, Beckenham, 1950.
16. Delacour, J., *Pheasants of the World*, London, 1951.
17. Woolley, Roger, *Modern Trout Fly Dressing*, London, 1932.
18. Williams, A. Courtney, *A Dictionary of Trout Flies*, 2nd edition, London, 1950.

CHAPTER VII

Shape

The hackle on a dry-fly has two functions to perform. First to support the fly on the surface of the water, and second to produce a good enough representation of the wings and legs of the newly hatched dun to fool the trout. There would appear to be a conflict here for the representation requires relatively few turns of hackle to achieve a good copy of the living fly, yet to keep the fly on the surface, the greater the number of turns the more hackle points there are to support the weight of the hook. A very good representation is not of much use if the fly becomes water-logged after one or two casts. On the other hand, a very buoyant fly is useless if it is so heavily hackled that the trout refuse to accept it. In a previous chapter, however, I have shown that even a fairly heavily hackled fly has not a great number of hackle points touching the water and the answer therefore depends on how heavily the fly can be hackled and still be accepted by the trout. For some reason, this appears to depend on the river being fished. The trout in some rivers are willing to accept a heavy hackle, but in others the practice is to fish a very lightly dressed fly and we are told that this is essential if trout are to be taken. Probably it has something to do with the speed of the river or the clarity of the water. A compromise must be the answer. Enough turns to make the fly fish right on the top of the surface and yet not so bushy that it will scare the trout.

Obviously to achieve this result the first essential is to use a really stiff hackle and this aspect has been dealt with in the chapter on Quality, but the shape of the hackle is also of considerable importance.

If we wish a fly which is not too heavily dressed and yet able to support the weight of the hook, it is easily seen that the hackle should have all its barbs of exactly the same length. If such a hackle was wound on to a hook, the barbs would form a circle of a fixed radius; the point of every barb on the bottom of the circumference would touch the water and would help to bear the weight of the fly. A barb which is shorter than the rest is useless for support, since it will not touch the water and the longer ones will be left to do all the work. To test the lengths of the barbs, hold the hackle by the tip with one hand and, gripping the barbs gently between the thumb and index finger of the other, run these fingers down the shaft so that the barbs are made to stand out on either side. If you repeat this once or twice the barbs will tend to hold this position. You may read in books that this makes the barbs stand out at right angles to the shaft, but in actual practice, this is not strictly correct. The barbs will spring back and take up an angle of about 70° to the shaft. If you try to force them to stand out at 90°, they will fracture at the base. The same thing happens when you wind a hackle on to the shank of a hook. If you look carefully, the barbs do not stand out at right angles, they are slightly inclined towards the direction in which the hackle is wound, but since it is wound round a wire of very small radius, the difference is not important. The ideal hackle, when the barbs are smoothed out, would produce a pattern where the tips of the barbs form a line parallel to the shaft. This when wound on to the hook will give us the circle that we require.

In a bad hackle, when the barbs are smoothed out, those at the base are longer than the ones further up the shaft, or frequently it can be the other way round, the barbs at the top being longer than those nearer the base. When wound, a hackle of either of these types will produce a poor fly, for the long barbs will be the only ones to touch the water, the shorter barbs doing nothing to aid buoyancy. The fault in the first type is further accentuated by the fact that the lower barbs have to bear the weight because they are longer and they are the least effective since they are always softer than the barbs higher up.

A number of writers have stressed the importance of choosing a hackle with points parallel to the shaft when spread and I do agree that this would be perfect if it were possible to obtain. Even so if you spread out a number of normal dry-fly hackles well shaped in the natural position, you will find that very few of them have a very bad shape when spread. In the illustration (Fig. 4) I have shown good and bad shaped hackles both in the natural position and when spread. I think that you will agree that the differences are less than would be expected.

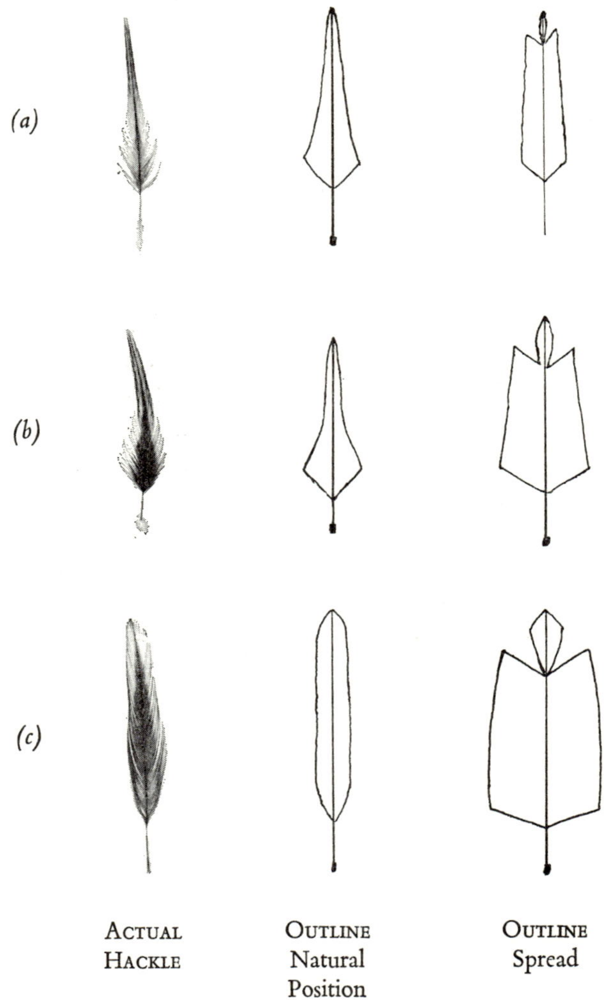
(a)
(b)
(c)
ACTUAL HACKLE
OUTLINE Natural Position
OUTLINE Spread

FIG. 4

Shape

It is normally said that the perfect hackle when removed from the cape is long, lance shaped and that the barbs should be short and all of equal length. It is assumed that all these are separate factors and that a hackle can have some without necessarily having the others. In fact, this is impossible. If a hackle has short barbs of equal length, it must be lance shaped or conversely, if the hackle is of a long lance shape, it must have short barbs of equal length.

The shaft runs right up the centre of the hackle, the point of the shaft forming the tip of the hackle. The barbs extend from either side of the central shaft, they are in pairs set opposite to each other, not alternately, and they are of equal length giving the hackle a balanced proportion unlike some feathers, flight feathers for example, where the barbs on one side of the shaft are longer than those on the other. The hackle, although it is very slightly curved convexly to conform to the shape of the bird's neck, is in one plane; the barbs, in a good hackle, lie against each other and even the tips of the barbs do not overlap the barbs above or below. Taking one side and counting from the top, the first barb lies tightly against the shaft. The second, if it is not to overlap the first, must take a wider angle to clear the base of the first and it then curves upwards to lie against it. The third has to extend outwards to clear two breadths of barb before it curves upwards and so it goes on, the hackle getting broader as it goes down. The point, however, is reached where the angle of the barb to the shaft does not need to increase in order to allow the barb to clear the one above. It is the length of the barbs which determines when this point is reached.

This is difficult to explain but I think that the diagrammatical drawing (Fig. 5) will make it clear. From point A downwards the angle of the barb to the shaft does not need to increase in order to allow the barb to clear the one above and the hackle would therefore remain the same breadth, but for another factor coming into operation. At a certain distance from the tip, barbules begin to appear on the barbs, increasing in number as you go down the shaft to form a widening triangular shaped web. The barbule formation depends on the quality of the hackle, a good hackle has few barbules, they do not appear until well down the shaft and they form only a narrow web. In a bad quality hackle the barbules start much higher up and they form quite a broad web by the time the base is reached. The start of the barbules and then the increasing number as they go down, makes each individual barb broader at the base. This forces each succeeding barb to take a wider angle to clear the barb above and so it goes on. This makes the outline of the hackle widen once again even although the barbs are still of the same length.

It is therefore the length of the barb and the formation of the barbules which governs the shape of the hackle. If the barbs are short in relation to the length, the hackle *must* be narrow and lance shaped. If the barbs are long, the hackle *must* be broad. Few barbules, giving a narrow web even at the base, and the hackle will be

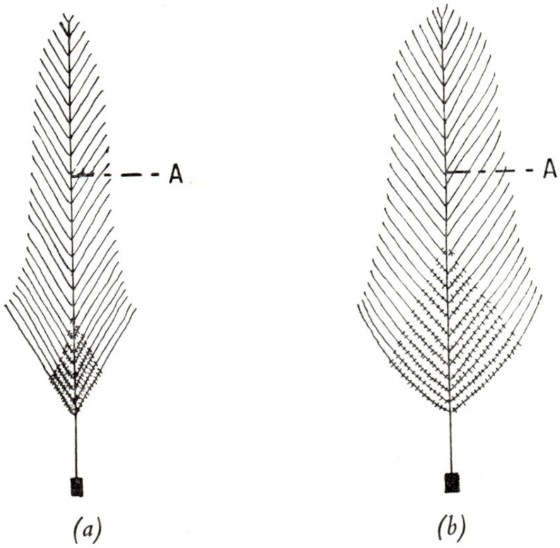

FIG. 5

narrow throughout its length. Barbules starting near the tip giving a broad web at the base and you have a triangular shaped hackle. By looking at a hackle you can tell whether the barbs are short or long and you can also judge the shape of the web. Fig. 5 diagrammatically shows what happens. *a* is identical in overall length and in the number of barbs to *b*, but the barbs are shorter in *a* and there is less web. It is the length of the barbs and web which governs the shape of a hackle.

I must admit to having a personal liking for a hackle shaped as Fig. 4 (*b*) on page 86, probably because this is the shape of a very good Red Jungle Cock hackle. The increase of barbules at the base causing the sudden broadening could be called a fault, and it might be thought better if the hackle were narrower throughout, but I have found that a hackle of this shape has that extra quality which is difficult to define, but which stands out. For choice I would always pick a hackle of this shape and it is this rather than extreme length and narrowness that I have aimed for in my breeding programme.

What happened to the barbs at the tip has always interested me. Were they naturally short so as not to extend beyond the tip of the shaft or was the shape of the top of the hackle produced by wear, the upper barbs being broken off? Using a high-power glass you can see that it is wear. A barb is broader at the base coming to a point; the barbs at the top, although narrower at the base than the barbs lower down, do not come to a point, they are broken off, so the shape of the tip is produced by wear.

It is fortunate that the wild Red Jungle Cock has a fine hackle, for in breeding it is easier to return to a character that is already present in the strain rather than to produce an abnormal one. Through the ages selection and cross breeding to produce birds of a large size and soft flesh or to produce a strain with a high egg-laying potential, has landed us with what we call soft feathered birds. The hackles on the cocks are broad with round tips and long soft barbs. Our aim is to return to the shape and quality of the wild bird.

There are two even more important reasons why a short barb is required in a good hackle.

The length of the barbs decides which hackle should be selected from the cape for a particular size of hook. For a given size of fly, some dressers prefer a longer barb than others and this also varies with the part of the country and the type of stream being fished, but to get the proportions right and to make a good representation of the living fly, there is not room for very great variation. In the standard hackled dry-fly, the length of the hackle standing out from the hook should be about the length of the hook shank, the shank being the part between the eye and the start of the bend.* In a winged dry-fly, the hackle length is usually rather less. Most dressers would, I think, agree with these proportions. It follows therefore that in each cape there are only a certain number of hackles which can be used for each size of fly. The normal range of hook sizes used for dry-flies is between sizes 18 and 12 (Redditch scale). This means that in a cape only hackles whose barbs are between 0·5 and 1·0 cm are of

[* Note by the author's son: The reference here is, I think, to the barb length i.e. the length of the hackle means the length of the barb when the hackle is wound on the hook.]

use. Good capes are rare and expensive. It is important therefore to have as many hackles as possible on the cape which are within this range. To understand what we are looking for, we must examine the make-up of a cock's cape.

On a cock's neck, the hackles overlap and are graded in size from small at the crest of the head to large at the base of the neck. Where the feathers at the top are small, there are a large number growing close together to cover the area, and at the bottom where they are large, there are fewer on the same area of skin. The table (Fig. 6)

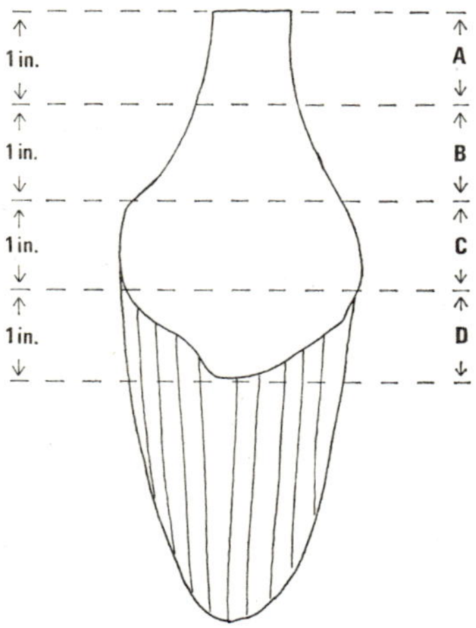

Note—1st year. Yellow badger cock of good shape.

Diagram shows cape viewed from rear.

Skin of this cape was exactly 4 inches long from top to base. It was cut at 1 inch intervals into 4 bands A–D. The hackles were removed from each portion of the cape, counted and measured.

It must be remembered that the shape of a cape can be greatly altered during drying depending on whether it is pulled lengthwise or outwards and this alters the area of each section. The figures given therefore are only a guide for a normal cape.

Total length of hackle		*Useful length*	*Length of barb*	*Ratio (see p. 92 for explanation of these terms)*	*No. of hackles in area*				*Total No. on cape*	*Size of hook (Redditch)*	*No. of hackles useful for hook sizes 12–18*
cm.	*inches (approx.)*	*cm.*	*cm.*		*A*	*B*	*C*	*D*			
under 1·5	½	—	0·4	—	261	8	—	—	269		
1·5–2·0	¾	1·3	0·5	2·6/1	89	5	—	—	94	18–14	345
2–3	¾–1¼	2·0	0·8	3/1	72	18	—	—	90		
3–4	1¼–1½	2·8	0·9	3/1	41	40	—	—	81	13	
4–5	1¾	3·8	1·0	3·8/1	21	59	—	—	80	12	
5–6	2	4·5	1·1	4/1	17	51	—	—	68	11	
6–7	2½	5·0	1·2	4·1/1	10	52	4	—	66		
7–8	3	5·8	1·3	4·5/1	3	42	8	—	53	10	
8–9	3¼	6·5	1·4	5/1	—	28	16	—	44	9	
9–10	3¾	7·4	1·6	4·6/1	—	15	25	—	40	8	
10–11	4	8·4	1·8	4·5/1	—	7	34	1	42		
11–12	4½	8·6	1·9	4·5/1	—	—	35	6	41		
12–13	5	9·0	2·2	4·1/1	—	—	5	29	34		
13–14	5¼	9·5	2·3	4·1/1	—	—	—	12	12		
Total no. of hackles in area					514 +	325 +	127 +	48 =	1014		

Fig. 6

shows the number taken from a cock's cape of good quality and good shape of hackle. This cape came from a first-year bird of moderate size. In a larger bird the total number of each size would be greater owing to the larger size of the cape, but the proportions of each size are approximately constant.

The first point of interest is the number of hackles in the useful range, i.e. 0·5 – 1·0 cm. In this cape there are 345 and they all come from the small band across the cape starting at $1/1\frac{1}{2}$ cm. down from the comb and ending at $3\frac{1}{2}/4$ cm. Fly-dressers with very nimble fingers can use hackles of a very small size, but the normal person finds it very hard to use a hackle smaller than 2 cm. ($\frac{3}{4}$ inch). As will be seen from the table the limiting factor for the number of useful hackles on the cape in the smallest size is the length of the hackle. The barbs can be of a useful length, but the hackle itself too small to work with. What we want therefore is a cape where the hackles at the top are long, but retain the short barbs. This increases the number of useful hackles in the smallest sizes. Similarly a cape with short barbs in relation to length increases the number of useful hackles before the barbs become too long for normal use and thus the 'band' which produces hackles useful for the normal size of hook becomes broader. If, on the other hand, the hackles themselves are short, they quickly become too long in the barb and the 'band' becomes very narrow. Long hackles with short barbs do therefore increase the number of hackles on the cape which can be used in the normal sizes of dry-fly.

Fly-tyers when looking for a good cape often say that they want a cape with a large number of small hackles for tying small flies. This really means nothing. What they want is a cape with long spear shaped hackles. Hackles of this shape have short barbs and a cape with such hackles will give them what they want.

From the practical point of view, there is a further reason why a long hackle is required. A dry-fly must have enough barbs to support it on the surface. Where the hackle is short and therefore has few barbs, two or even three hackles may be required to produce enough barbs, where one or at the most two long hackles would suffice. Whether this is an important point depends on the fly-tyer's point of view. Using more hackles makes it more difficult to make a neat fly and from a professional's angle it must slow down production.

The ideal hackle is therefore a very long spear-shaped feather and this shape always has short barbs. A cape with feathers of this shape will have a large number of hackles of the sizes required for dry-flies and the long hackle will make tying easier.

A word of warning. Not infrequently, particularly in the dark colours, you may come across a hackle which at first sight looks long, narrow and lance shaped, but if it is examined it is soon noticed that each barb is very soft, fine at the point and actually quite long. Instead of running straight out from the shaft and ending, the tip of this soft barb turns upwards taking the line of the outside edge of the hackle and lying on top of or underneath the tip of the barb above. The result is a hackle which appears to be of excellent shape, but a hackle of this type is quickly spotted if you know what to look for, because the fine ends to the barbs running up the line of the outside produce a dark edge to the hackle. Wherever this dark edge is seen, the hackle is suspect and a second look will show whether it is indeed of this poor quality.

It is easy to say that a hackle should be short in the barb and long in length, but without a standard for comparison such remarks mean very little. What is long and what is short? It was John Henderson,[1] in his article to the *Fly-Fishers' Journal* in 1932, who produced what I consider to be a neat, easy and accurate method of evaluating quality of shape. When tying a fly the whole of a hackle cannot be used; the downy part at the base must be stripped off leaving what can be called the 'useful' length of the hackle (Fig. 7). If the length of a barb in the centre of the hackle is measured, the ratio between the useful length and the length of the barb gives a very good guide as to the standard. Anything better than 3/1 is, in my opinion, reasonable, 3·5/1 good, and if you have a hackle of over 4/1 then you are in the very good class. This rating refers to hackles in the sizes useful for dry-flies. In the larger sizes, the ratio improves so that 4/1 is not uncommon. The table of measurements of hackles from a good cape shows the typical pattern. The best hackle for shape that I have ever seen came from a wild Red Jungle Cock. The hackle had a useful length of 3·6 cm ($1\frac{3}{8}$ inches), barb 0·6 cm ($\frac{1}{4}$ inch), a ratio of 6/1, which is quite outstanding in this size. John Henderson took his useful length as the distance from the bottom barb to the top barb which could be tied in after allowing for the tip to be held in the hackle pliers. This is of course strictly more accurate and will give a slightly lower ratio, but I find it easier to measure the length right to the tip rather than stroke back the top barbs, and the difference in results is small. It is I think advisable to use the useful length, the length after the downy portion has been stripped off rather than the total length, for hackles can vary quite considerably in the length of the part that has to be removed and it is the ratio of useful length/barb that matters. The useful length does also depend on the fly-dresser's opinion as to how many of the bottom barbs should be removed, one tyer being more particular than another, but again the difference would be small.

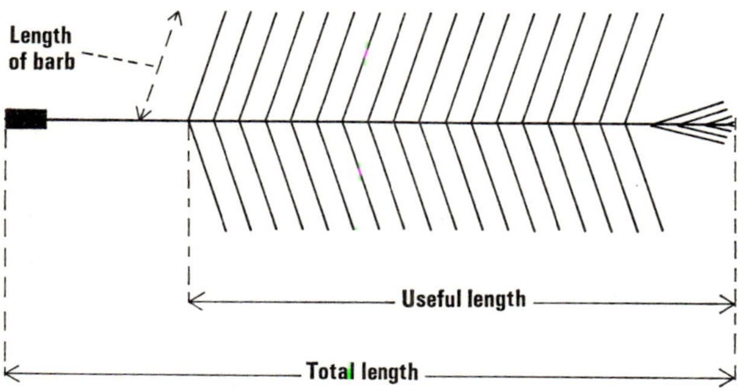

FIG. 7

Total Length	8·0 cm.	Ratio	=	Useful Length	/	Length of Barb
Useful Length	6·0 cm.		=	6·0	/	1·3
Length of Barb	1·3 cm.		=	4·6	/	1·0

Note—In practice it is not really necessary to spread the hackle in order to measure the length of a barb in the centre.

This ratio invented by John Henderson gives a simple and reliable method of judging a hackle for shape and I think that it is a pity that it is not universally used. Jack Heddon has suggested that instead of writing the useful length/barb length as a ratio it should be expressed as a number which he has called the Hackle Index (H.I.), i.e. 3·5/1 is a H.I. of 3·5.

I would like to draw attention to the differences in the shape of the down at the base of the hackle. In the normal breeds of poultry this down does not extend far up the shaft, but is broad and fluffy. The down at the base of the hackle from a Red Jungle Cock is different; it extends further up the shaft than normal, but it is sparser and not nearly as broad; it is a good deal narrower than the true barbs. I like to see a hackle with this shape of down since it normally indicates that the hackle itself is of high quality.

What have the experts in the past had to say about the shape of hackles? Halford[2] went into the subject in great detail and made a horrible mess of it. I am sure that he must have deeply regretted having put it down on paper, for his error has been pointed out over and over again. He described the perfect hackle and drew a diagram to show what its shape should be. After the soft barbs at the base had been removed and the barbs were spread out, it should be the shape of a geometrical triangle, the barbs at the base being longest and reducing regularly in length to the tip. He even went into such detail as saying that the length of the barb at the base should be one-quarter of the useful length, i.e. a ratio of 4/1. If it was 4/1 at the base and the barb lengths were to reduce regularly it would have a ratio of 10 or 12/1 at the centre of the hackle and much more by the tip. How he could write in this way is hard to explain; he must have written it down without really thinking what he was saying or checking against a real hackle. A hackle with the shape of an isosceles triangle that he suggests is quite impossible to obtain in nature; it could in theory taper regularly from base to top when in the natural position, indeed a very fine hackle is not far from this shape, but no hackle could be of this shape when the barbs are spread out.

Pulman[3] in his third edition of 1851, speaking of cock hackles, favours the pointed shape and condemns hackles with 'a shape broader towards the point than at the quill end. To make a good fly with such a hackle is next to impossibility.' As he differentiates in the requirements for a hen's and a cock's hackle, it would appear that here he is speaking of a cock's hackle for a dry not a wet-fly.

Dr Cutcliffe,[4] who knew so much about hackles, does not seem to put great emphasis on the shape although he says the 'shape of the feather should be an isosceles triangle, having its base at that end which is inserted into the skin, and its elongated apex, slenderly, gradually and evenly tapering off to a fine point – many will be found suddenly ending in a rounded extremity, these are not so good, but some of them if of fine colour may be retained'. He was obviously speaking of a hackle in its natural shape, not spread out. His shape therefore is what we like today.

McClelland,[5] the practical fly-dresser, says that 'with regard to the shape of hackles, I personally am not too hard to please. . . . For dry flies it is certainly desirable that hackles should be long in the quill and relatively short in the fibre, though this may be obviated by using two hackles for one fly.' He goes on to mention Halford's perfect hackle but considers that for ordinary flies hackled at the shoulder only 'the advantage of such feathers would be expressed by a minus quantity'.

Stoddart[6] writing in 1847 considered that shape and quality were of great importance in a hackle. He stated that 'The selection of the hackle, then, requires considerable care and knowledge. Not one cock in ten walking the farmyard, yields feathers of this description, truly available to the angler. They are generally found to be stiff and long in the fibre – seldom prettily tapered – and, when colour is brought into consideration, perpetually at a discount. The annexed is a sample of what may be reckoned, in point of shape, a good trouting hackle.' His drawing depicts what I would call a very fine shape for a cock's hackle, in fact it would be impossible to produce a better hackle. This in 1847! There is no mention of a hen's hackle and yet he goes on to say that the feather from a plover's crest or from the neck of a starling is excellent for small flies. It is difficult to decide whether he really likes a stiff or a soft barb. By 1857, Stewart[7] states that a hen's hackle, because of its soft barbs, is much to be preferred to a cock's hackle.

Leonard West[8] in his book of 1921 has a coloured plate of hackles and the shapes of these are what we today would call very bad indeed. Presumably he chose his examples with some care, which means that he must have considered shape of little importance and he barely mentions it in his text.

Skues[9] considered that of the three factors colour, quality, shape, the third was the least important for he said, 'if colour and condition be right, you can do with something less than ideal shape'. Dr Baigent was of the same opinion. In his private notes he wrote, 'Hackles of good quality are long and taper slowly and gradually from base to point. This gives the maximum number of plumes or fibres of equal length. This point has been much dwelt upon by some writers and given undue importance. In making of palmer flies it is important, but for the dressing of most others there are sufficient plumes of most ordinary hackles of equal length or approximately so to complete the fly. Where this is not so two may easily be used.' I have examined a number of the capes which Dr Baigent had taken from cocks of his own breeding and which he had used to provide the samples for his book. The colours I admired tremendously, magnificent capes in the rusty blues and blue reds which he liked so much, but the shape of the hackles would not be considered good today.

As no one before John Henderson used the ratio figure, there are no figures to go by and there are few accurate drawings of hackles in angling literature from which we can judge shape, but I have examined quite a large number of hackles of 1870–1900 era and in all cases I would call the shape very bad. I have a number of hackles, for example, from birds bred in the 1870s by the famous Welsh fisherman Griffith Lewis. Both the colour and the quality are very good, but the shape is very bad with a ratio of about 2·7/1. Fortunately we have the wonderful collection of fly-tying materials in the Flyfishers' Club in London, which I have already referred to in my chapter on Colour. The shape shown in a number of cases is good without being outstanding and while I know that the best that I have been able to breed is, as regards shape, better than anything in this collection, I also consider that the average imported cape available at the present time, is, judging from this collection, of better shape than was available to fly-tyers in those days.

It seems clear to me that while good shape was known to be an asset, it was not until the 1930s that the majority of fly-dressers began to consider it to be of importance. There would be some who sought for perfection, but the majority like McClelland, Baigent and Skues put it as the least important of the three qualities. While we all look for perfection and want colour, quality and shape all to be perfect, I think that in this they were correct. The colour must be what is wanted and unless the barbs are stiff, the hackle is useless for a dry-fly. A hackle with barbs of the correct length must be chosen, but if the hackle itself is not long enough to give enough turns, a second or even a third hackle can be tied in. Extra work, and it takes more care to produce a neat fly, but despite what Stoddart said, I do not think that it is too difficult for a good dresser. I know that fly-tyers have different ideas and that very light dressing is the normal practice on certain rivers, but I must say that I myself rarely use only one hackle. I normally use two to produce a really buoyant fly. I therefore agree with Skues that if necessary 'you can do with something less than ideal shape', but the emphasis is on 'if necessary'.

A hackle of bad shape offends the eye and there are also fewer hackles of usable size on the cape. From a breeder's point of view there is one good thing. Good shape and stiffness tend to go together, provided the shape is really good and not of the type which at first glance looks good, but has in fact long soft barbs turning up at the point. By breeding for shape, we also breed for quality.

What has been the opinion of writers in more recent times? Harding[10] in 1931 favours the slow taper. Woolley,[11] another very practical dresser, asks for a hackle which is 'long and fine in the stem, the fibres short in proportion'. In America, Cross[12] in the late 1930s knew a good shape and he drew attention to the bad type I have mentioned where the points curve upwards. He states that this type only seems to appear in the dark colours, black and dark red, and I agree with this. Jennings[13] also gave the modern idea of a good shape. Leiser[14] in his book on fly-tying materials gave a long description of what he considered to be a good hackle. He was obviously speaking mainly of capes imported from the East. Although he has a very wide knowledge of hackles, for some reason he made the same mistake as Halford for he produced drawings showing good and bad dry-fly hackles and they are of the impossible Christmas tree shape.

Commander Walker,[15] as one would expect, gave a very clear and accurate description of what a good shape should be. Veniard[16] agrees with the modern idea, but makes the statement that 'A hackle from an old bird will have much shorter fibres in comparison to its length than will that of a hackle from a young bird'. He gives no facts or examples to support this assumption. It has often been stated that the barbs in an old bird are stiffer than in a young bird, a statement that I have already discussed, but this is the only place where I have found it stated that the barbs are actually shorter in an old bird. In my experience of breeding, I have no evidence to support this statement. I can note no difference in ratio of useful length/barb in hackles taken from the same cock over a number of years.

References

1. Henderson, John, 'Some Notes on Blue Game Fowls and Hackles', *Fly-fishers' Journal*, Vol. 21, Spring 1932.

2. Halford, Frederic M., *Floating Flies and How to Dress Them*, New York, 1886.
3. Pulman, G. P. R., *The Vade-Mecum of Fly-Fishing for Trout*, 3rd edition, London, 1851.
4. Cutcliffe, H. C., *The Art of Trout Fishing on Rapid Streams*, South Molton, 1863.
5. McClelland, H. C., *The Trout Fly Dressers' Cabinet of Devices . . .*, 1898; 7th edition, London, 1931.
6. Stoddart, Thomas Tod, *The Angler's Companion . . .*, Edinburgh, 1847.
7. Stewart, W. C., *The Practical Angler*, Edinburgh, 1857.
8. West, Leonard, *The Natural Trout Fly and its Imitation*, 2nd edition, Liverpool, 1921.
9. 'V. C.' [Skues, G. E. M.], *Silk, Fur and Feather . . .*, Beckenham, 1950.
10. Harding, E. W., *The Flyfisher and the Trout's Point of View*, London, 1931.
11. Woolley, Roger, *Modern Trout Fly Dressing*, London, 1932.
12. Cross, Reuben R., *The Complete Fly Tier*, U.S.A., 1936; reprinted 1971.
13. Jennings, Preston J., *A Book of Trout Flies*, New York, 1935; reprinted 1971.
14. Leiser, Eric, *Fly-Tying Materials*, U.S.A., 1973.
15. Walker, C. F., *Fly-Tying as an Art*, London, 1957.
16. Veniard, John, *Fly Dressers' Guide*, 2nd edition, London, 1966.

CHAPTER VIII

Hen hackles, saddles and spades

Unlike cock hackles, which are used for dry-flies where a hackle with stiff barbs is required to keep the fly sitting right on the top of the water, a hen's hackle is used for wet-flies where the very opposite is wanted. Hen hackles have not evolved as display feathers; their use is functional. They have not developed into feathers with stiff shiny barbs, and normally have barbules right up the barbs. The barbs themselves are soft, so the whole feather is excellent for holding the water. This allows the fly to sink immediately it lands on the surface and the soft barbs move in the current giving a good imitation of the legs of the nymph. Shape and quality are not very important; colour is the main point of interest to the fly-tyer.

Broadly speaking, the colours of hen hackles are the same as found in the cock, but there is a difference. In many hen capes, the hackles are edged with a different colour. This you do not get in the cock; if there are two colours, the other colour is in a well defined position as a list in a furnace or over the feather as in a rusty dun. It does not appear as an edge as is often found in the hen hackle. Hen capes where the feathers are edged in this manner are often extremely beautiful, particularly in the blues.

The basic colours are very similar to those of cock hackles, except that the red is a much yellower red; the fiery red of the game cock is never found.

In single colours, white is common; ginger is not as often found as red, and the red is a brown-red. Black is easily obtained.

In bi-colours, badger is not uncommon, but in this colour, particularly in the game breeds, the shaft and a thin strip outside it is often white, not black, so you have a white centre, black list and white outside. The black list is frequently more of a grey than a good black and not uncommonly it can be brown, producing a badger with a brown list, something that I have never seen in a cock.

The outside colour in the hen hackle is white or near white. I have never bred a hen with the yellow of the yellow badger.

The furnace in the hen is also different from that of the cock. You never get the true bright red of the cock, the outside colour being brown or ginger. The list is frequently more of a grey than a good black and the white shaft marking is also common. I do not think that you ever get in the hens the black showing at the base of the feather only as you do in the cock. In the hen furnace, more often than in any other colour, you can obtain a hackle which has barbs not unlike those of the cock, the barbs lacking barbules on the outer half, and I have seen some which could very nearly be used for a dry-fly. The shape also in this colour can be much longer and narrower than in the normal hen, but it has a spoon shaped, not pointed, tip and although narrow for a hen, never achieves the spear shape of the cock hackle.

In barred, the colours are the same, cuckoo and creel, but I do not think that I have ever seen a creel with black. In the hen hackle, the bars appear to be more indefinite and blurred than in the cock.

Duns. For some reason it is much easier to breed blue dun hens than it is to breed blue dun cocks. I always seem to have a high percentage of good pale, medium and dark hens, but never enough good cocks in this colour. The blue dun hen capes are a very beautiful soft blue, almost a lavender blue in the medium and pale shades. Rusty dun is unknown in a hen. The hackle is never overlaid or suffused with red or brown as in the cock. Hen honey duns are very beautiful capes and they are not difficult to breed. The blue list is normally a very good pale to medium blue and well marked. Pale to medium blue seems to be the normal colour for the list and this is what is wanted; there is not the same high percentage of very dark, near black, which unfortunately appears so often in the cock. The outer colour is a good pale, medium or dark honey.

In addition to those basic colours, many capes have every hackle, or sometimes only those at the sides of the cape, edged with a different colour. Black hackles can be edged with gold, producing a most attractive cape, or with white. White can be edged with black, although I have never seen this in game. The blue duns, particularly in the pale to medium, can have the hackles edged with gold and these capes are very beautiful.

Saddle Hackles

Saddle hackles are the upper tail coverts, which in the cock have become elongated and hang down on either side of the tail. These are very beautiful hackles, the colours being very strong and clear, and in the bi-colours they are very well marked. You can always identify a saddle hackle from a long neck hackle by the shaft. Presumably because they are designed to droop down and not remain stiff and straight, the shaft of the saddle, particularly near the base, is much thinner and softer than that of the neck hackle. If you strip the down away from the base of a saddle, you often break the shaft. The barbs, unfortunately, are too long to allow it to be used for a normal size of dry-fly. They can, however, be used to tie a very large fly such as is required for dapping. They are also very good for streamer and salmon flies. The colour of the bird seems to influence the length and shape of the saddle hackles. I have found that the reds in particular have much longer and narrower saddles than those produced by birds of other colours. Blue dun saddles never seem to be so long.

Barbs from saddle hackles are often used for whisks, but I do not think that they are nearly as good as the barbs produced by spade hackles for this purpose.

Spade Hackles

Spade hackles come from the back of the cock. They take their name from the shape of the hackle, which is like a round-ended digging spade. Strictly, the spade hackle comes only from the scapulars which are the bunch of feathers on the back which cover the gap between the back and the wing, but the medium coverts from the wing are often called spades. Spades are very beautiful hackles. I wish that neck hackles could be obtained in the very beautiful colours, markings and shades that appear on spades. The barbs are very long and stiff and they are ideal for whisks. It has been suggested that the whisk produced by a spade hackle is too stiff and that it can prevent the fly from going into the trout's mouth. I find this very hard to believe.

CHAPTER IX

The commercial supply

In the early days of fly-tying, the supply of hackles would come from the farm yard, but as the colour and quality became more and more important to the discriminating tyer, the search would widen. There has always been a small number of enthusiastic fly-dressers, both amateur and professional, who have kept their own flocks of poultry to provide themselves with hackles, but their number has never been large. It has always been carried out as a hobby and the supply of hackles from this source has always been unimportant. Towards the end of the nineteenth century, as the number of fly dressers increased, the butcher's shop and the big poultry markets became the source of hackle supply for the growing number of professional fly-tyers and also for the businesses which were established at that time to supply the fly-tying materials required by both amateurs and professionals all over the country.

If this book is to give the whole history of the hackle, the story of the commercial supply for fly-tyers should be properly covered. I have been exceedingly fortunate in that Mr Sam Harris has given me the fascinating inside story of the build-up of this trade.

Sam Harris started business in London in the 1930s as a foreign exchange dealer in a merchant bank, but the economic depression at that time made him look for other employment. He joined a small firm of feather merchants in London and quickly found that feathers of all types fascinated him. His firm obtained feathers from poultry and game shops and he visited the fishing tackle shops, selling them the feathers that they required. About 1936 he joined Paul Glaser and they founded their own business of wholesale feather merchants at Luton. Hackles and feathers for fishing flies form only a small part of their business, but even in this line they are easily the largest and most important wholesale dealers in Britain, so there is no one who can give us a more accurate account of the supply of hackles.

I asked him to write the story of the commercial supply of hackles since the 1930s and it is so interesting and informative that I cannot do better than quote verbatim the whole article that he has written for me.

In writing about the commercial aspect of the hackle business, it is impossible to divorce it from the subject of feathers in general. If one wished to concentrate solely upon hackles, which I suppose is possible, it would be found that prices would have to be paid beyond the capacity of the manufacturer to absorb.

The early origin of fishing flies is a subject for researchers far more competent than I. But there is little doubt that fly-dressers, professional and amateur alike, used to rely on the local butcher for their supply of hackles. There were also dozens of men who called on poultry and game shops to pick up items like rabbit and hare skins, pheasant tails, pheasant, game and wildfowl plumage and wings, and so on, and then sell them to skin and feather merchants. The latter began to receive enquiries for fly-tying items; the most important were hackles, all varieties, and they in turn passed on this information to the shop collectors, who acted accordingly.

Between the wars, both Leadenhall and Smithfield markets became important sources of supply. Indulgent bosses allowed some of their staff to take feathers off birds being dressed for sale, to earn a few extra shillings a week. Some bosses shared the proceeds, but most did not. The saving of feathers spread far afield and important centres were Newcastle and Birmingham, the nearest big towns to Alnwick and Redditch. Up to that time, I had not heard of many heads of poultry being cut for capes, at least not on a commercial basis. The small firm of Messeena, of Clapton, London, and later Leamington Spa, used to obtain small quantities of capes from local butchers, until on the death of the owner the business was taken over by Veniards. A number of amateur tyers have told me they used to get their own capes cut locally, but at no time were English capes a commercial proposition. Indeed, it was found that the big manufacturers and the dozens of small professional tyers in Redditch and other towns without exception preferred to buy their hackles put up in paper. A cock (or hen) was stripped of the long hackles first, which were laid across approximately 7 inches of newspaper, the medium hackles placed on top, and the tiny beak hackles (16/20), the most important, were placed on top of them. The paper then represented a complete head of poultry. It was folded approximately 7 × 4 inches but this varied according to the person preparing the papers.

The speed with which these packets were put up was phenomenal. If a head is stripped of hackles by means of a blunt knife immediately after slaughter, the feathers simply peel off. When our business was situated not 200 yards from St. Paul's in London, and therefore in the shadow of Smithfield, hackle papers could be bought a gross at a time at a price something like one old penny each! No premium was put on colour: the sorting was left to the merchant. There was the occasional exception. One of the Leadenhall shop employees told me he had actually sold a single pure natural blue dun hackle (one feather!) for 7s 6d to an enthusiast who saw it on the bird in the shop. It was the merchant's job to ensure that the birds were not hackled during the moult, and immediately after, when the feathers were growing anew. This aspect of hackles is more important than the age of the bird. Some of the finest hackle packets I ever handled came from a Norfolk Packing Station from its Christmas pluckings. Therefore the birds were unlikely to have been older than 9 months, considered the maximum for a table roasting bird.

As far as Britain is concerned, there has been a vast change over the past ten years or so. The big tackle manufacturers have to all intents and purposes closed the fly-dressing departments. Allcocks, and to a lesser extent Hardy's, no longer tie flies. Milwards discontinued to do so even before being absorbed. Indeed if these firms wished to tie flies, it would be found well-nigh impossible to get the workers, who daily diminish in numbers. The age of the machine does not brook the competition of any craft industry.

Nevertheless, the number of fly anglers is on the increase all the time. The chief way to obtain flies is by dressing them themselves, which most of them do. There is a great danger here, however. I would not dispute the fact that the practised amateur is the best fly-dresser of all. But there are among them a number who, seeing the result of their expertise, want to set up making flies in a professional capacity. Unfortunately, however, many are not business men. Their enthusiasm prevents them counting that most precious commodity of all: TIME. It is like growing one's own tomatoes. If one reckons the time spent on them, they cost 50p per lb, but we justify this by saying, 'Well, I grew them myself.' The object of making flies professionally is to make money. Many start up a fly-tying business and almost as many fail. I have told many a potential customer not to make flies for a living. Few listen.

However, the demand for materials, particularly of hackles, develops constantly, as the popularity of the sport takes hold. There is no thrill like that of catching a trout on one of one's own flies. In the States, there is a greater chance of success for the fly-tyer, because a great number of dressers work from their own sports shops, from where flies are sold to the public, thus eliminating distributors' profits. The good amateur dresser is much more demanding than the old professional firms. Fred Hardy, that prince among fly-anglers, once told me at Alnwick in answer to my question, it wasn't so much a question of what fly to use, provided one knew how to fish. He said all this with a twinkle in his eye, for he had a great sense of humour. Trout are far more fastidious than salmon in taking a fly. Speaking of salmon, the dictum of a Deeside ghillie should always be remembered: 'If they're gaun tae tak' they'll tak' the handle of a chanty' (a chanty being a good old-fashioned chamber pot!). I have this on the authority of Norman Halkett of Huntly, Aberdeenshire.

Indeed, Redditch rarely complained of quality as between dry-fly and wet-fly hackles, provided price was reasonable and types well defined. But then, between wars and for a few years after the 1939 war, 80% of flies seemed to be wet-flies. The amateur has changed all that. He now wants not only dry-fly capes, but top quality too.

Now I believe I know dry-fly hackles as well as most, but I have perforce to look at the question with a different eye. I have to study *availability*. I cannot tell you when the first capes were cut, nor do I believe anyone else can. Certain it is that between the wars professional firms bought and preferred the necks put up in papers, as I have already described. Here and there one would get hold of an English cape. It would appear that the earliest production of capes on a *commercial* scale originated in America, from which country I used to purchase thousands of white, red, badger, Plymouth Rock, and a variety of grizzle called a variant, which was usually slightly tinged with ginger. The quality of white was never good, as they were cut from Wyandottes, which produce wide and softish hackles, well suited for wet-flies. The Plymouth Rock capes could on occasion be absolutely superb.

But let me take you across the sea to Poland. At the beginning of the century and even before then, at a place called Mielec in eastern Poland on the Russian border, there was a community of people who lived and breathed feathers. During the severe winters they experienced, the time was passed grading and sorting fancy and bedding feathers for the various world markets. This place was the nursery of the feather merchant. Many of their number emigrated, some to Paris, others to New York. Among them was a man, Charles Zucker, whom I consider one of the finest judges of feathers in the world. He took over his father's business in Paris, called 'Les Fils de B. Zucker', and was the man who gave me my first chance by allowing me several thousand pounds worth of credit, back in 1935, when I started up as a feather merchant myself, having served my 'apprenticeship' for a short period with a small London firm. From then on, for many years, except during the war itself, I used to go to Paris 11 or 12 times a year, often returning the same evening. Of course, I had contact with a number of firms there.

Zucker had his own French representative in China, and he used to buy sizeable lots of fancy feathers in Shanghai and other great centres. His brother, Nathan Zucker, was already established in New York, and the two of them did a vast business, Charles acting on Nathan's instructions from New York. Labour costs in America were extremely high, and ways were sought to compensate for this by getting other countries to do some of the work.

Thus it was that Charles Zucker (or his representative) taught the Chinese to string hackles and many other types of feathers. This was pure exploitation. The Chinese graduated the larger hackles from 2½ or 3 inches up to 6 or 7 inches long and passed a needle through the hackle quill, thus joining them with thread to produce lengths of hackles around 4 feet long, which they do to this day both in neck and saddle hackles. American users of feathers, particularly fancy feathers, will not consider them unless strung, enabling their own labour force to peel off and use hackles and the like, without having to select feathers beforehand. This of course did not apply to the small hackles (3 inches and under) used in fly-dressing, and in which we deal extensively. But the large strung hackles are to this day by far the best proposition for sea jigs and seatrout flies. America used vast quantities of strung saddle hackles, the longer the better, for streamers, and although it is not generally known in England, certain good qualities of saddles are used to make dry-fly streamers in the States. They must be narrow, pointed and fairly stiff.

In 1939, Charles Zucker, then in Paris, was called to the French army. On the capitulation of France, he escaped to the south, and then, via Spain, found his way to New York, where the firm he established there does big business to this day, although *not* mainly in tackle feathers.

After the war, it was Zucker who asked Chinese merchants (not yet Government controlled) to go into the question of producing capes. He was not yet to see the fruits of his labour, since *Red* China had no trade agreement with America for some years. However, these American merchants were not to be so easily beaten. They descended upon Hong Kong and got the local dealers to string feathers (hackles and many other types) for the American market. To this day, Americans send out quantities of unsorted feathers to Hong Kong to be sorted, sized and strung in Macao and Hong Kong itself. Some quantities of capes came on the market via Hong Kong, but of Red Chinese origin. The first capes were excellent but gradually a picture was built up of quality according to origin. Hong Kong and Cantonese capes are on the poor side. The capes from Shanghai are little better. Tientsin necks are much better, but still not good enough.

The fault with Tientsin capes lies in the cutting. For some reason the neck end is cut in a squat manner, too wide, giving the impression of loss of beak hackles.

The one place in China which produces beautifully cut capes is Tsingtao in Shangtung Province, from which, if available, first, second and third qualities are, or rather were, obtainable. I suppose, over the years, I have bought from Tsingtao alone, over 500,000 capes, but for around 2 years I have failed to obtain an offer (except for fourth quality only, which I turned down). This situation obtained since America made a trade agreement with China in 1972/3. Between the end of 1971 (when I went personally to Canton) and now, the price of capes has risen 7 or 8 times what they were, in part accounted for by the 36% devaluation of sterling (at the time of writing). One cannot bargain with the Chinese, who nevertheless are totally honest. Nor could one hope to buy specific items like capes unless a number of other items are bought in combination. For this purpose, one has to have an interest in millinery, bedding, badminton and other feathers also. A letter of credit has to be opened and often it is four months before one sees the goods.

It is getting increasingly difficult to make the sale of Chinese capes pay. First one always had to buy first, second and third qualities (Chinese grading, which incidentally, would not suit the English market) in equal quantities. An approximate breakdown of colours of Chinese capes is:

	%
Red	26
Light red	6
Chocolate	2
Ginger	4
Pale ginger	1
Black/red (not furnace)	12
Furnace, cochy bondhu Greenwell	3
Cree	12
Grizzled	4
Badger	4
Light Sussex	4
Natural black	2
Nondescript	6
White, cream	9
Rejects	4·9
Duns	0·1
	100

The exigencies of the 'do it yourself' trade in materials has meant for some years the rarer colours are those in most demand. Thus, at present we have around 70,000 reds in stock, 25,000 black/reds and 10,000 crees. In other words, although our good colours are not exactly cheap in wholesale terms, the return for them cannot compensate for the poor sale of the commoner colours. From 1960 to 1972 this did not matter much, the capes in bulk being so cheap. But with prices rocketing 7 or even 8 times since America was able to enter the Chinese market, the lack of sale of reds, crees etc. is a serious matter for the importers. A fly dressed with a red hackle is sold for as much as one dressed with a badger. Yet the original cost is vastly different.

This aspect of bulk buying in capes, is, I think, the most disturbing. The wholesale buyers of capes can hardly pay more than they do, and yet if they do not, supplies could dry up.

The Americans are a law unto themselves. First of all, tackle firms themselves do not buy in the Chinese markets. The obtaining of supplies is left to the feather merchant. The latter is guided by demand and lack or abundance of supplies of a given commodity like capes. It is not uncommon in the States for prices to rocket double overnight. Alternatively, if the shipping lists, which are eagerly scanned by the merchants, disclose the approach to U.S. waters of a big quantity of a hitherto scarce item, the price can fall by as much as one-half. These days the Americans seem to be looking more to India for the supply of capes.

Capes are being produced in fair quantities there. Many of the Indian capes are of fair quality, albeit rather small and therefore deficient in feathers. There is a fair proportion of dry-fly capes, coming all of them from gamecock birds, a great number of which are wild, tree-resting types. Most of these capes find their way to the States, where insistent as the importers are to have only the best, they leave the secondary qualities for other countries. The great defect in Indian capes is the high proportion of nondescript colours, but the main trouble is with the exporters themselves, many of whom are small, poor firms eager to eke out a precarious living, and seldom reaching the impeccable commercial standing of the Chinese. I would mention in passing that Taiwan is a country to be avoided as regards feathers. Samples are *never* representative from that country.

But to return once more to Mielec, a town responsible for more international feather dealers than any other district. Among the most famous of the families is one with the name of Topiol. I first met Michel Topiol, then a young man of about 20, when he first arrived in Paris, and where he had set up a sorting atelier, near the centre, where a staff of women sorted goose and duck feathers impeccably. We did business with him from the very beginning. Turning to the subject of fishing trade hackles, I discovered that

Topiol's father, still in Mielec, was prepared to sort out small stiff hackles (say 1 to 3 inches) in white, red, badger, furnace, grizzle etc. These proved to be of superlative quality and we imported quite a quantity until the old chap was swallowed up in the maelstrom of 1939.

Incidentally Michel Topiol continued in France, avoiding the Germans and establishing himself in the feather and down trade after the war. Today, he has a factory on the northern outskirts of Paris, at Argenteuil (of Claude Monet fame), as large as the Middlesex Hospital, and it is the most famous of the bedding firms in Europe. He is a multi-millionaire, and although we do business with him to this day, he is on a vastly different plane from us. I just wanted to show what the feather world owes to the town of Mielec. Other centres in Poland were situated at Cracow, Katowice, Warsaw, Poznan and Kielce. From these centres the Government body called Animex which deals in food preparations and raw materials like feathers, still holds sway, but the main interest is bedding. Why this is, is a long story, and would not help in our research on the subject of hackles.

France itself never produced hackles for fishing. In any case, that country has hardly the tackle history of England, despite that Mouches-Ragot is probably a bigger manufacturer of flies than any British firm today. The French disliked the hackle papers, but bought such American capes as were offered avidly. Norway, a great centre of fly-tying activity (Mustads) bought the cheaper colours of papered capes in quantity, and as many hens as cocks, if not more, signifying the manufacture of great quantities of wet lake flies.

Italy was a great producer of hackles, which, however, were not of use for flies, as practically all cockerels were caponised, rendering the small hackles wide and soft.

Denmark became of interest. I well remember being approached by the Co-operative of Ringsted offering to supply us Greenwell (ginger furnace) hackles, which were so strange that I did not recognise the form. Since big quantities of other types of feathers were involved, I went over there. To my amazement I found that the manager was plucking the tiny feathers found on the birds' knee joint! He had a goldmine in his hands and did not know it. The time was around 1947. Denmark produced thousands and thousands of pure-bred Brown Leghorn hens, and had farms of which this bird was the sole type. Taking capes was out of the question. The business was too extensive to permit the slowing up by cutting capes. But I got them to strip off the small hen hackles (each and every one a superb Greenwell) and bought them by the kilo, 50 kilos at a time. The wings of this fabulous bird yield the brown speckled 'Alder' quill and I confess I sold some hundreds of pounds at a give-away price. If only I had known! Strange to relate, the Brown Leghorn cock does not yield a Greenwell head, but rather a ginger. With broiler production in full swing over there, the Brown Leghorns are now non-existent. It cannot have escaped notice that practically all broilers are white. It is for the same reason that over 99% of turkeys are now white, and that the brown or khaki Campbell duck has disappeared. The trade in broilers demands the white flesh so necessary for freezing.

Holland assumes some importance for the fly-dresser, but only in so far as wild fowl feathers are concerned. There is no production of *fishing* hackle feathers. Belgium produces no hackles at all and Germany produces no feathers of note.

Jugoslavia, Hungary and Poland produce only goose and duck sortings. Poland was a great supplier of long creamy hackles in pre-war days, putting them up beautifully in great hunks of hackles, bound securely and evenly with string. But the only small hackles came from old Topiol.

It is possible to buy supplies of capes from Kenya, where good tuition in taking capes was given by Ted Crawford of the Kenya Fishing Flies Co. of Kericho, now closed down. Mr Crawford is now with Hi-Flies International of South Africa. Much of this business has to be taken on trust. I would not impugn the honesty of those offering hackles, but I do rather doubt their ability to pass judgement. I believe a business is possible with capes from the Philippines, but I have no first-hand knowledge at present.

One must not forget the taking of grey jungle cock capes in India, which started around 1920. Hackle capes came much later.

I cannot let all this pass without reference to an experience I had with an Irish Packing Station. Immediately after the war, in the search for feather supplies in Eire, I was introduced to the Castlebar Bacon Co. in County Mayo. The poultry side was run by an Englishman, E. A. King of Hampshire. He was very keen to do business with us.

There, in slatted crates I saw hundreds and hundreds of head of poultry, half of them hens, it is true, but there were whites, Greenwells, badgers, Wickhams, ginger, game cocks, Black Minorcas, and a good sprinkling of duns. In short an old-fashioned collection of types the like of which it would be impossible to find in Europe or America today. We must have imported tens of thousands of hackles put up in papers, one head per paper. Then suddenly came the disaster of the wet pluck. We could not understand how King managed to continue the business with apparent unconcern, for we knew that wet-plucked birds were frozen for marketing. This meant that in order to keep the flesh white, the birds had to be bled upon slaughter, to prevent congealed blood accumulating on the breast. The birds would become bespattered with blood, which rendered the hackles useless. Eventually King came to England when this type of poultry business stopped, broilers by then having increasingly intruded upon the scene, so that breeding was no longer of any commercial importance.

It seems that he had, to use his own words, hit on the idea of injecting a tiny quantity of chloroform in the ear of each bird, which then went drowsily to sleep. He then had the hackles stripped off, after which the birds were slaughtered. He assured me that this was all with the blessing of the local agricultural inspector, but I was personally sceptical of it.

We collected hackle packets from a great number of packing stations in our time. Some excellent specimens came from Scottish packers. All this business has completely finished. I know of no packer who produces feathers other than white in this country, which means broilers of no practical use.

In America, one is still able on occasion to buy a few hundred Plymouth Rock capes, produced by individuals raising the birds for the sake of the capes. Previously tiny P.R. cock hackles from $\frac{1}{2}$ to 2 inches long, of superlative quality, could be bought. Neither America nor any European country, except England and Norway, ever took kindly to hackles put up in papers.

Nevertheless the cape only came into its own after the war, and particularly so from about 25 years back. I am speaking commercially, of course. Capes were cut almost from the time flies were dressed, but only in very small numbers for private individuals.

What of the future? I am afraid I am by nature a pessimist, mainly because I think the optimist is a menace, who can only look to be disappointed, whereas the pessimist can only be heartened if his pessimism is unjustified. In my humble opinion, the greatest bar to abundant hackle supplies in England is the United States. The dealers there search ruthlessly for what they want, and if they have to pay more to get it, more they will pay. We enjoyed a good deal of success in the years 1950 to 1972 simply because the States were not permitted to deal with the People's Government of China. We are still, to some extent, living on the fat of this good period for England. In fits and starts it may yet come back, but I am not counting on it.

The hackle trade will, in my view, have increasingly to rely on Indian capes. I do not like the prospect, but I feel a way round all this may yet be found. I am today sending a cable once again to China for capes, but can almost anticipate the one word reply . . . 'unavailable'!

CHAPTER X

Breeding for hackles

As I have said in the first chapter, when I decided to breed poultry for hackles, I was amazed to find how little information there was on this subject. Thomas Hughes,[1] John Henderson[2] and Captain Evans[3] are really the only people in this country who have given us any information to work upon and the American books that I have read do not carry the subject very much further. Guy Plas breeds cocks for hackles in France and I understand that he has written a number of articles on the subject. These should be studied, for some of his hackles are very fine. I now realise that breeding for hackles is a very complicated subject. After nearly fifteen years of breeding I appreciate that I have only scratched the surface and I cannot tell anyone how it should be done. All I can do is to describe in some detail the methods that I have used so that these notes may be of some help to anyone who wishes to try this fascinating business.

All the angling books said that the Andalusian was the cock to produce the steely blue hackles, but it appeared to be accepted that the finest of all hackles were produced by the Old English Game cock, the fighting cock of the pit, and so I decided to try to obtain a few of this breed.

The Old English Game cock has an air of romanticism surrounding it, for cock fighting is one of the oldest sports in the world. The game cock was, according to some authorities, brought to Britain by the Romans and the sport flourished until the early part of the nineteenth century, when it was prohibited by Act of Parliament. It, like horse racing, has been the sport of kings and Charles II in particular was very attracted to the pit. To this day cock fighting is carried out in many parts of the world. In the seventeenth and eighteenth centuries, when the sport was perhaps at its height, the number of fighting cocks kept in Britain was enormous. Bringing the cocks up to their highest pitch of fitness for a fight was a great art, and expert handlers were employed to look after the birds. Good fighting strains were carefully guarded and feeding methods were kept a close secret, for much money could change hands at a main.

The story of cock fighting is most interesting and one which should be studied, for while one may dislike the sport, over the years breeding for the pit became a great art and in the literature there may be some information which could help anyone interested in breeding for hackles. Breeding for the pit and breeding for hackles have a good deal in common, for the aim of both is to produce a cock with hard, tight feathering. A cock with soft loose feathers is of no use in a fight and for fly-tying the hackles are much too soft. It happens therefore that the cocks selected for their fighting qualities had hackles which suited the fly-dresser and it is probably this continuous selection for the pit which, over the centuries, produced a bird with fine stiff hackles.

It is, however, interesting to note that it was not until the second half of the nineteenth century, when the dry-fly purist held sway, that the demand for stiff hackles arose and by that time, cock fighting having been made illegal, there could not have been nearly as many birds available as in earlier years, although no doubt the sport still flourished extensively behind closed doors. (Even today cock fighting is by no means dead in Britain.) Although cock fighting may not have been openly carried out in the days of Halford as in earlier years, the last half of the nineteenth century was the great period for poultry exhibitions. Poultry of all kinds were kept in large numbers for the show bench and the Old English Game was no doubt one of the most popular breeds. There would be a good supply of hackles from this source.

There are I am sure many breeds of game cocks in the world, but in this country there are three distinct species: the Old English Game, normally referred to as O.E.G.; the Indian Game or Asil; and the Modern Game. The Indian is a thick-set, heavy bird, which is used extensively for cock fighting in the East. Atkinson[4] mentions a number of different types of this eastern bird, which are no doubt common in their own area, but in Britain, I have only seen the Indian Game on the show bench. The Modern Game is a strange looking bird, the result of very selective breeding to produce a new type for exhibition. It has a small body and very long legs and is to my eye a very unattractive bird. I have never examined the hackles from the Indian or the Modern very carefully, but the shape of these birds does not look to me as if they would provide good hackles for our purpose, the only possibility being that the Modern Game might in certain cases produce a cock with a very well shaped hackle. Atkinson also mentions the Spanish Game, but I have never managed to see a cock of this breed.

It would, I am sure, be possible to build up a strain of cocks with fine hackles from a number of breeds. Poultry can be divided into two main groups, hard and soft feathered. In the hard, the feathers lie close to the body; in the soft, the feathers are fluffy and stand out from the body. None of the soft breeds produce hackles which are of any use for dry-flies, but there are a number of breeds of hard feathered birds, which could produce good hackles. The Leghorn is an example. It is tight feathered and it can be obtained in white, black, blue and brown. This breed has been used to produce hackles by some breeders, including Reuben Cross.[5] In other parts of the world there must be many local breeds which would be excellent for the purpose. The O.E.G. cock, however, is a breed which we know to be capable, through selective breeding, of producing hackles of very high quality and shape; it is a hardy bird which can be kept easily in this country and it has the final advantage of occurring in every colour and marking known in poultry. From this one breed therefore we can obtain white, ginger, red, black, blue, badger, furnace, creel and cuckoo. With all these good points it would appear sensible to use this breed, but if anyone was in a part of the world where the O.E.G. could not be obtained, I am sure that he would be able to find a breed that was suitable and would produce a very fine strain.

In the days of cock fighting, by selection for the pit, there may have been cocks which had hackles which were perfect for dry-flies, but as we have seen, there are three essentials for a fine dry-fly hackle, colour, shape, quality, and to be first class it would have to have all three characters. When I started to breed O.E.G. I quickly found that today's birds certainly did not come up to this standard and instead of breeding purely to obtain the colours that I required, I had to select for all three characters. A very, very much bigger task.

When breeding to obtain a certain character, it is essential to have that character somewhere in the pedigree of the animal. In horse racing tremendous prices are paid for stallions and mares that have done outstandingly well, for the breeder knows that these characters can be passed on to their offspring. In cows a high milk production strain commands high prices. It is the same with poultry, breeding for colour, shape and quality; these characters must be in the strain or they will never appear.

When I started, the shape produced by my cocks was very bad indeed. The barbs were long in relation to the length of the hackle and I decided that this was the first thing that must be improved, but where was I to obtain a cock with a really fine shape of hackle? In his article in the *Fishing Gazette* of 1922, Thomas Hughes[1] mentioned that he had seen a bird called a Yokohama, which had a particularly long and pointed neck hackle. The Yokohama is a very extraordinary breed of poultry. It was evolved in Japan and was, I believe, so called in this country because the birds that were sent here came from that port. I don't think that the experts have quite decided what happens in a Yokohama, but in Japan cocks with tail feathers 20 feet long have been recorded. They are kept in special cages so that the feathers do not get broken. In this country the tails of the Yokohama are long compared to the normal cock, but there must be a secret which we do not know, because they never achieve the length known in the East. Once again I was very fortunate, for Mr Langton, who gave me a tremendous amount of help and advice when I started to breed, gave me the name of Mr McEwan of Lumsden, Aberdeen, who kept this breed. I obtained a beautiful pair of Yokohamas from him. The shape of the hackle of the cock bird was extremely good; very long, pointed and short in the barb. It was, however, rather softer than I would have liked. The Yokohama is not unlike a brown-breasted red O.E.G., but the shape of the body is different, being a long bird, while the game should be broad-breasted and short in the body. The bare skin over the ears of the Yokohama is white while in the true O.E.G. the ears are red.

I crossed these Yokohama with my game and immediately obtained a big improvement in hackle shape. I had produced a cross breed and although I kept breeding back with true O.E.G., even to this day the mixed blood shows up as my birds have white or some white on their ears. I am afraid that I would never win on a show bench.

The wild Red Jungle Fowl produces the nearly perfect dry-fly hackle for shape and stiffness. (The colour of course is always red.) It has only one very bad point in that the colour of the underside of the hackle is a chalky white. All hackles tend to have this chalky colour on the reverse, but it is most apparent in their ancestor the Red Jungle Fowl. However, the shape and quality is so good that I decided to use this breed also. Again I was fortunate in finding someone in England who kept these birds and he kindly sent me a setting of eggs from which I hatched a few birds. The cock is like a small black-breasted red game and the hen is a very drab partridge colour. I found that they cross-bred without difficulty with my O.E.G., using either the cock Jungle Fowl with game hens or vice versa. In the resulting chicks there was no sign of infertility. Indeed Delacour[6] mentions that in Britain today it is very difficult to obtain a pure species of Red Jungle Fowl, so many having domestic blood in their parentage. Again I noted an improvement not only in shape, but also in quality. Now that I knew that I had improved both quality and shape, I considered that it was a question of breeding and selecting to produce a strain which showed both these qualities to an acceptable standard.

I quickly realised that if I was to obtain a good strain of cocks within a reasonable number of years, I would have to hatch and raise a large number of chicks so that I would have enough young birds from which I could select my breeding stock for the next year. I therefore started five to six breeding pens, which gave me enough eggs to hatch some three hundred chicks. These young birds were kept on until the cocks had grown their adult plumage, which is about eight months after hatching, but it can vary by a month either way. When they had obtained their full plumage they were carefully examined in order to decide which would be kept for next year's breeding. The only point that was taken into consideration was the quality and the shape of the hackle. The general shape or size of my birds was never considered, another reason why my birds could never go on exhibition.

Particularly in the first few years, it was amazing to see the variation in the hackles. You wondered where some of the cocks had come from with such soft broad hackles. Their capes were not even worth keeping. Others, however, were good and each year I had the few which were better than anything that I had produced before.

This yearly culling is vitally important in my method of selective breeding, therefore a short description of how I carried it out may be of interest. As I will explain later, it is almost certain that unless you are dealing with just a few birds, you will be handling these cocks at night. I purchased a set of metal leg rings, each with a separate number, which are easily fixed round the leg. The birds were caught and every bird was ringed. At the same time about six hackles were plucked from its neck and placed in an envelope numbered according to the ring. With the aid of one helper, I could have a sample of every cock's hackle in about a couple of hours. In daylight the hackles were laid out and assessed. Normally about 75% could be discarded immediately as not worth keeping and then the 25% were re-examined very carefully. I normally did my ringing and first sampling a month or more before the birds had completed their growth of adult hackles and, as I have said in an earlier chapter, it is most unfortunate for a breeder for hackles that the neck hackles appear to be the last feathers to complete their growth. You must therefore be careful when selecting hackles at this stage not to take any that still have any of the sheath remaining, for it is not easy to gauge the quality and shape of a hackle until it has finished growing. The good point about this early collection is that it allows you to concentrate on the cocks that may be of use and possibly to kill off immediately birds which are really poor. You may also find that one or two are far more fully feathered than the rest and that their hackles are indeed fully grown. These can be killed off if not in the top class – feeding birds needlessly is an expensive business. This speed of feathering is I understand a hereditary factor, so having established your strain, only early featherers could be used for breeding.

I am also speaking as if all the birds were hatched on the same day. This is of course ideal, but you will probably have a flock of young birds of mixed ages – in my case ringing and sampling was carried out at fortnightly intervals.

Eventually you will have decided on the cocks to be kept for the next year's breeding; the others are killed off and their capes taken. Always compare the hackles from the cocks that you bred the year before with those from your young birds so that you are sure that you really have got something better.

I have spoken only of selecting the best cocks; what about the hens? Hughes[1] says that the cock plays the greater part in the colour of the chicks and the hen on the quality of the hackle. Unfortunately, he gives no facts to support this statement, but coming from him it should be seriously considered. If this is true, the selection of your breeding hens is vitally important when looking for improved quality, but I do not know how you can judge from a hen's hackle as to whether it will produce a good quality cock. The Yokohama hens do have longer hackles than the game hen and this is a point which could be watched. My selection of hens for breeding has been more on a group rather than on an individual bird basis. For one thing, since I run at least eight hens with each cock, I would need to raise a tremendous number of birds if I were to be as selective as I have been with the cocks, which in turn would mean that I would have to keep more hens to provide the hatching eggs for next year and so it would go on. Once my hens are fully grown and feathered – this happens very much more quickly than with the cock – I select them during the day as they run in a flock. Any 'mossy', that is loose-feathered, birds are netted and removed. Once I have reduced the flock to nice compact tight-feathered birds, my final selection is done on the colours I wish to keep.

Everything that I have said to date refers to breeding for quality and shape, but there would be little purpose in breeding poultry for hackles, if the rare colours, the mystic blues, were not the final aim.

Reuben Cross,[5] the American fly-dresser and writer, bred cocks for hackles and states that he produced blues by crossing white × black. This is certainly possible in theory and may not be difficult to do, because the 'blue' in poultry is not really blue at all but a grey formed by black pigment in a proportion of the cells, the grey ranging from near black to near white according to the proportion of black and white cells. I have a feeling, however, that in either his black or his white there was a blue ancestor.

You can try the black × white method and you may quickly obtain success. I have not tried it. I do feel, however, that you save yourself a lot of work and possible disappointment if you manage to obtain a blue cock. A blue of any breed is difficult to obtain and this has always been the case.

The trouble with blues is that they will not breed true. The books say that according to the Mendel theory, blue × blue will produce 25% black, 50% blue, 25% splashed. (Splashed is the name normally used for this bird. It can be practically or completely white or more normally it is a bird which is mainly white with a greater or lesser proportion of black streaks. In some examples, the colour can be a pale grey with black markings, but the base colour lacks the brown overtone of the true blue dun.) 25–50–25% is the book percentage, but I am sure that there is something in the makeup of the blue fowl which is not yet understood, for with me in cocks at least 10% blues is nearer the correct figure. Breeding blue hens is not nearly so difficult and 50% blue hens might not be far off the mark. There has been very little work done on colours in poultry, because it is of no commercial interest. Lippincott[7] does, however, note that blues which breed 100% blue would appear to be possible, but it would require a very detailed breeding programme using one cock and only one hen, followed by a careful analysis of the strain produced, so that if a true breeding blue did appear, it would be picked out. From this bird a strain breeding pure blue could in theory be built up.

My advice therefore is to find a blue O.E.G. or a blue in whatever breed you have decided to use. If you can't get a blue in the correct breed, I would use a blue of any breed in order to get that colour into the strain.

Let us suppose that you have tracked down a source of blue O.E.G. You will almost certainly be disappointed in the colour of the cock. The hen will probably be a nice medium blue with, if you are lucky, a certain amount of gold in the hackle particularly on the side of the neck, but the cock, while a lovely pearl grey on the breast, will almost certainly have hackles which are difficult to tell from black. It could have some gold or red over the back, which is not a fault. The shape and quality of the hackles are certain to be very poor. You have to settle down to improving these characters while retaining the colour. Everyone who has bred for hackles eventually realises that it is in the reds that the quality appears and this is the source that you have to use.

In order to obtain enough material for selection, I have bred in quantity. I hatch my eggs in an incubator and when hatching some hundred eggs at a time, I have found no simple method which will allow me to identify the chick coming from any marked egg. I have therefore been unable to follow the results obtained from any one pen. I have worked on a mass selection basis.

While selecting the cocks for shape and quality of hackle, I have not been completely unmindful of colour. Except for some special reason, I have used nothing but blue hens or hens bred from blue and these are mated with my selected cocks. In addition and running parallel with this programme of selection entirely for quality and shape, I keep one or two pens going with blues only. From these I select the best of the blue cocks and it was very encouraging when the time came when some of my blue cocks were amongst the best of any colour.

Breeding blue × blue you get this high proportion of black and splashed cocks and I have found that both these colours tend to have hackles of poor quality. This is indeed unfortunate for according to Mendel's theory black bred from blue × splashed will give you 100% blue. If you can get the quality and shape in black or splashed this crossing should be tried.

If you breed your blue dun cock, with its nearly black hackle, with blue hens, you will find that any blue cocks that you do produce will have hackles like the male parent. I have found that the best answer to this is to include a number of splashed hens in the breeding pen and, using this method, cocks with medium/pale blue dun hackles can be produced.

If anyone were taking up this fascinating hobby of breeding poultry for hackles, I again stress that they ought to study the article by Hughes[1] in the 1922 *Fishing Gazette*. He hatched all his chicks under hens and was therefore able to keep a careful record of the results obtained. This I have been unable to do, but it is the only method by which theories can be tested and an accurate record of results can be compiled. I cannot really add anything to the knotty subject of how to breed blues.

Going through the colours used by fly-dressers, the following points gained from my experience may be of use to a future breeder.

Self colours, except for the blues, are fairly simple. To obtain white, I would cross splashed × splashed and I am sure that you would obtain a good percentage of pure white. Quality might, however, be your difficulty. My gingers have appeared and I am not very sure of the parentage. They are a very useful hackle and if I was trying to produce them, I think that I would try crossing a red cock which had as little as possible black at the base of the hackle × splashed. If you ensured that the red had a well shaped, high quality hackle with very little or no black at the base, then I would expect some nice gingers to appear, although you might obtain a fairly high proportion of splashed which contained red as well as black markings. These three-colour birds are not uncommon even when breeding blue × blue if there is red in the blue parentage. Once gingers are obtained I would expect it to be fairly simple to build up a strain which bred true. Dark reds are, I think, best obtained by crossing a good red × dark blue rather than by using black. The dark red obtained from blue is the blue red favoured by Cutcliffe and Baigent.

In a good bi-colour, the list should run right to the tip of the smallest hackle. This can be obtained by selection and I do not think that it would be difficult to produce a strain that always had perfect markings, for a poorly marked bi-colour is of little value. You should only use cocks with well defined lists, but you must take care not to have the list too broad. The true badger, as compared to the yellow badger, I have found rare, but if required it would not be difficult to raise a true breeding strain using a grey cock. The grey is a handsome bird with grey not blue breast and a white hackle with black list. The yellow badger is a very common colour when breeding blues and using reds to improve quality. Strictly speaking they are not badgers at all, for the list is in fact a very, very dark blue not black. To be exact they should really be called very, very dark blue honey duns with yellow rather than honey outer colour. If you look at the large hackles at the base of the cape, you will see that the list is indeed blue not black. These birds always have blue breasts. The quality of the hackles is usually very good and frequently the tips of the barbs are black.

The furnace should I think be selected and bred purely for this colour. Starting with a black-breasted red cock of good hackle quality and shape and with the black list as well marked as possible, crossing this bird with Partridge hens, you will by selection be able to produce a furnace with perfect list in the small hackles. I have never tried to produce a pale brown red furnace, but I do not think that it would be difficult to lighten the colour and once it was obtained, it would not be difficult to fix a strain. Hughes maintained that the coch y bonddu, a furnace with black tips to the barbs, was very difficult to produce, the only method being to breed furnace × black, but he emphasised that the black must be a true black and not bred from blue. Personally I have not found a furnace with black tips to the barbs difficult to breed, nor have I found it essential to use a true black. Black hens bred from blue when crossed with furnace have given good results, but the quality can be variable.

The horizontal bars are interesting. The true O.E.G. cuckoo or creel is now a rare bird and I have never seen one. Atkinson[4] gives creel, creols, mackerels etc. as local names for cuckoo and describes the bird as having black or dark blue bars on a pale ground, similar in colour to a Plymouth Rock or Cuckoo Dorking. He does not mention the bird where the black is replaced by red although I am sure that this must be an omission for it is the colour which is normally called creel, cuckoo always being black on white. In creels there are frequently a number of black bars along with the red giving a very attractive hackle. The cuckoo is much in demand in the south of Scotland for the Border series of flies Grey-Hen and Yellow, Grey-Hen and Rusty etc. and, being unable to obtain a cuckoo O.E.G., I experimented to see whether I could introduce this colour into my strain. I obtained a very well marked Plymouth Rock cock and crossed it with blue O.E.G. hens. The results were very successful. I find that the cuckoo marking is very persistent and although I have never used another outside bird, after eight years I have a strain with black and white bars which breed true to colour. I also have at times cocks which are true creel colour, some with red and white only and others with red, black and white. The persistent barring also appears to a greater or lesser extent in cocks of all colours giving a most attractive grizzle hackle. Blue dun with an indication of barring is a most useful hackle.

I have raised my chicks almost entirely in incubators. I may have been unfortunate or it may be bad management, but I have found that with O.E.G. eggs it is difficult to obtain a hatch of over 60%, which is most disappointing. It is not due to infertility for the ones that do not hatch either just 'pip' the shell and get no further or are dead but fully formed in the shell. I have had eggs examined in the laboratory, but the scientists cannot produce the answer. The eggs appear to have a much thicker shell than those of a normal fowl and I am of the opinion that it is a question of providing the correct humidity at the time of hatching. Pheasant eggs are difficult to hatch in incubators and require high humidity during the last few days and I feel that it is probably the same difficulty with O.E.G. When the hen is allowed to hatch her own eggs, I have found the results variable; sometimes she takes away a good proportion of chicks, at others only two or three.

John Henderson[2] in 1932 stated that he had given up incubators in favour of natural hatching because he had been told by Skues that the latter produced better hackles. In his article in 1949, however, he indicated that he was not now certain that this was correct. Personally I can see no possible reason how the hackle quality, which is a hereditary characteristic, can be influenced by the method of hatching.

The chicks when hatched are delightful little birds in all possible colours, the blues in particular being most attractive. They are not difficult to rear.

If, as I do, you are keeping your young birds inside on an intensive system, it is important to have two sheds and to separate the young cocks as soon as you can identify their sex. This is not easy in the immature birds; the lengthening saddle hackles or the curved tail of the cock is often the first sign. I am sure, however, that anyone who studied this question would soon be able to spot the differences at a much earlier age. It can be done by experts at the day-old stage and this must be the correct method to adopt for efficient management.

Fighting amongst the young cocks I have not found to be a serious problem; provided they are raised as a flock they soon develop an order of precedence, but a stranger introduced in daylight is another matter and, in consequence, any movement should be done at night. A backward or ailing bird is harried unmercifully and unless of particularly fine colour these birds are best culled on the spot, for even if kept separate and given every attention, they never develop into good specimens with good hackles.

Whether de-beaking has, or has not, been carried out at a young age, as the young cocks begin to develop their adult plumage feather pecking becomes a very serious problem. Gripping another cock's hackles with the beak becomes a favourite pastime. The hackles gripped are either pulled out or are frayed, bent or broken and are quite useless. It means destroying the bird or keeping it for a year until the next moult. I have tried hard de-beaking, keeping the birds in near darkness – which helps – and a number of other methods, but none are 100% satisfactory. Within the last few years a special beak ring has been developed which I have found to be the answer to the problem. No doubt this ring will be improved, but as made at present, it is a wire ring covered with plastic. The ring is flattened at the bottom and there is a gap at the top leaving two open ends. The ring is slipped over the upper mandible, the open ends clip into the nostrils and the ring lying between the top and bottom bills prevents the beak from closing completely. It does not affect the feeding of the bird in any way, but does not allow the beak to grip a hackle. As well as beak ringing, I recommend that the young cocks should be fairly heavily de-beaked. There are simple electric machines which will allow this to be done quickly. All this sounds rather drastic, but it does not seem to upset their feeding. I would suggest that all your young birds should be de-beaked and ringed at the stage where they cease to be chicks and become immature birds. Choose the correct size of ring, and afterwards go over the flock frequently to ensure that none have lost their rings. At the same time, with a pair of nail clippers, trim back the edges of the beak if it is growing round the ring. This can happen amazingly quickly, allowing the bill to grip. Feather pecking is the greatest danger in intensive hackle production and should be prevented at all costs. A year's work can be ruined in a day. With a small number of birds running free range, feather pecking rarely occurs although I have seen it happen in the open.

I have already described my method of looking over the cocks, selecting and culling as the hackles become fully grown until I am left with only my breeding stock for the following year.

It is almost certain that good dry-fly hackles will become more and more difficult to obtain. The industrialisation of even the more backward parts of the world is proceeding at high speed, and efficiency in poultry production is the death of quality hackles. You can breed for egg production, for poultry meat production and for hackle production, but the higher the efficiency in egg or meat production, the poorer the hackles. It may be that the perfect artificial hackle will be produced. I find it hard to believe that this could happen, and if it did it would be at great cost. Just as we cannot produce anything to equal a spider's thread, so I cannot see that the barb of a fine cock's hackle could be manufactured.

Some day, I believe that someone will settle down to produce high-quality dry-fly hackles in quantity, otherwise future generations will not know what it is to fish a dry-fly as it should be fished.

The O.E.G. will certainly be kept alive in Britain for many years to come and I would advise anyone to start with this breed. I think that I was also correct in obtaining a Red Jungle Fowl to give the shape and quality. Although the wild bird is not as common as in the past, I am sure that it will be obtainable from zoos. Blue I think you must also obtain and this may be one of the greatest difficulties, for blue is not popular on the show bench. I would proceed as I have done, breeding in quantity, to obtain enough birds for selection, for I am sure that this is the only way to obtain results reasonably quickly.

Having obtained quality and shape to the standard required, you have to settle down to find the answer to breeding for colour. My first plan would be to find out if it is possible to tell the adult colour from the colour of the chick. Wing tagging with numbered tags at the day-old stage, with careful cataloguing of the chicks' coloration and markings, might enable one to predict with some certainty the colour of the adult. Unwanted colours could then be culled at the day-old stage, and with sexing at hatching, unwanted pullets could also be culled. It is the cost of raising birds which are ultimately of little value that runs away with the money. Chick colour to adult colour has never been properly studied. John Henderson[2] did a certain amount in this and his notes should be read, but I have not found his results to be accurate with my birds.

Then there is the problem of colour. I feel sure that there must be an answer to this. Having obtained the quality that you desire, I would then settle down to build up pure bred lines in the principal colours, for I cannot see how you would work out a breeding-for-colour programme, when you have a collection of birds which have a variety of colours in their blood. Pure bred lines means in-breeding and this is treated with horror by some biologists, but it would appear to have been normal practice in the days of cock fighting. The fighting cock had to be in the peak of health, strength and stamina when put to the test. If in-breeding produced a weakly cock then the cockers would certainly have known this. Lewis Wright[8] quotes from the notes of John Douglas who used to breed for the Duke of Newcastle and it is clear that he in-bred to the extent of father × daughter and son × daughter for twenty years or more. He states that it is very important to keep very close strains if the colour is to be kept pure. It is clear from Sketchley[9] that close breeding was a normal practice and he cites a case where a pure line of black reds suddenly produced two piles. On checking their back history, there had been a cock of this colour used twenty-one years before. Sketchley was also very insistent that a cock of over two years should never be used for breeding or the strain would quickly deteriorate. While carrying out this in-breeding they would of course watch for the slightest sign of weakness in any bird and it would immediately be destroyed. It would be this most careful selection which would allow them to maintain a perfect strain.

It is not often realised how great was the number of game cocks kept in this country. Almost every noble family had their particular strain of cocks and there were also hundreds of 'cockers' or 'gentlemen of the sod' as followers of the sport were called. Atkinson[4] says that one breeder alone had three thousand cocks at walk, each cock's pedigree and parents on both sides being known. He goes on to say, 'Many people imagined that these birds were of all colours, but it was not so: the different colours were bred much more perfectly than at present, and eye, beak and leg must match, so that, although there might perhaps be a hundred cocks or more of one strain in the pens for training, yet they would be almost all alike to a feather.'

Every cocker seems to have specialised in a particular colour. Atkinson mentions Lord Derby's Black-breasted Reds, Lord Hill's Silver Duckwing, Mr Nunis' Yellow Birchens, Captain Mansell's Piles, Lord de Vere's Blacks and on the duns he says, 'There are no more celebrated strains of fowls than the various strains of dun. Sir Francis Boynton, Mr Elwes (one of whose red duns won the extraordinary number of twenty-seven battles), and that great cocker and soldier, Colonel Mellish, bred some of the finest ever seen. They are particularly hard and workmanlike in appearance, and the hens have a neat, trim and gamey look.' He states that the duns are valuable for artificial fly-making.

I cannot believe that the breeders had to cull out half or more than half of their stock to produce these hundreds, if not thousands, of dun cocks all identical in colour. The secret of breeding dun must have been known. Has it been lost or is it recorded somewhere in a little-known book? Can a fly-dresser imagine three thousand blue dun cocks? Can it be done again?

References

1. Hughes, Thomas, 'How to Breed Fowls for Fly Tying', *Fishing Gazette*, 1922. 'Breeding Game Fowls for Fly Tying', *Salmon and Trout*, 1928. 'Feathers for

Artificial Flies', *Feathered World*, 1928; reprinted in *Fishing Gazette*, 1928, and *Flyfishers' Journal*, Vol. 21, Spring 1932.

2. Henderson, John, 'Some Notes on Blue Game Fowls and Hackles', *Flyfishers' Journal*, Vol. 21, Spring 1932. 'The Old English Game Fowl and its Breeding for Hackles', *Fishing Gazette*, 1949. 'Hackles and Feathers of the Game Fowl', *Fishing Gazette*, 1949. 'Retrospect' (On the Breeding of Old English Game Fowls for their Hackles), *Flyfishers' Journal*, Vol. 5, Autumn 1962.
3. Evans, J. D. D., 'Further Notes on Blue Game Fowls', *Flyfishers' Journal*, Vol. 22, Spring 1933.
4. Atkinson, Herbert, *The Old English Game Fowl*, London, 1891.
5. Cross, Reuben R., *The Complete Fly Tier*, U.S.A., 1936; reprinted 1971.
6. Delacour, J., *Pheasants of the World*, London, 1951.
7. Lippincott, William A., 'The Case of the Blue Andalusian', *American Naturalist*, Vol. 52, 1918.
8. Wright, Lewis, *The Illustrated Book of Poultry*, London, 1890.
9. Sketchley, W. Gent, *The Cocker*, 1814; reprinted Weymouth, 1968.

CHAPTER XI

Preparing the cape

A short note on preparation is included for completeness. Whether you have the whole bird or whether you have the head and neck alone, after the bird has been plucked and dressed, the method is the same.

Lay the head, beak towards you, comb downwards. With a sharp pointed knife or scissors make an incision through the skin of the throat below the beak. Using a strong pair of scissors, cut through the skin from this incision down the edge of the hackles till the bottom hackles are reached; continue round the neck until you again come to the throat; up the edge of the hackles on the other side to meet the original incision. Be careful, when cutting round the neck, to cut below the bottom hackles. Rather than cut right round the neck, I find it easier to cut half-way round from each side, the cuts meeting at the back. Strong scissors are required for the skin of the neck is thick.

Changing to a small pair of pointed scissors, turn the head till the beak is pointing away from you. Cut up the side of the hackles towards the beak, above the eye until the comb is reached; continue round the back edge of the comb and down the other side to meet the original incision. The skin of the neck is now quite clear of the flesh and the two can be separated.

Grip the comb with the fingers of the one hand and gripping the hackles behind the comb with the thumb and first finger of the other, peel the skin away from the skull. A firm but steady pull will bring it away cleanly, but *do not* pull it too far down while holding the top of the cape. The membrane of the ears attaches the skin to the skull and if you attempt to remove the whole cape by pulling from the tip, it is almost certain that the skin will rip across at this point, which spoils the cape. Pull firmly, but not strongly until a resistance is felt, this means that you have reached the ears. Still holding the comb or head in the one hand, drop your grip with thumb and first finger of the other to the point where the membrane of the ears is attached. Pinch hard with this hand and pulling firmly on the head with the other, the cape will be pulled clear of the ears without damage. Continue with a steady pull until the cape is clear of the neck. You will find that the cape naturally folds very neatly in on itself feather side out, and if you have done it properly there is no blood-staining or damage to any of the hackles.

The cape has now to be cleaned and dried. The normal method of killing a fowl is by 'wringing its neck'. This is actually a quick and neat method of breaking the neck behind the skull, causing instant death, and from a poultry point of view is an ideal method, for it breaks the blood vessels and the blood collects in the neck leaving a white flesh. From our point of view, it leaves the blood exactly where we do not want it. However, it can be coped with and I prefer this method of killing to one where blood may stain the hackles.

Spread a good thickness of newspapers on a table and have a packet of borax powder at hand. Domestic borax, not the purified medical borax, is not expensive. Lay the doubled cape feather side down, top away from you, on the newspaper and open it out. You will find that the skin is covered with a rather nasty mess of blood and membrane. Sprinkle liberally with borax all over. Grip the bottom left-hand corner firmly with the left hand and with the right, starting from the same point and dusting with borax as fresh membrane appears, rip the membrane away from the skin, working up and across from the starting-point. The membrane can be very firmly attached to the skin and a good pull is required. If done properly, the whole of the membrane from the bottom to the area of the ears can be removed in one without much mess. Use plenty of borax, which dries up the blood. If only the one hand is used to rip away the membrane, the other, holding the cape, is kept clean and the hackles are not soiled. The membrane above the ears never comes away with the rest and it has to be removed in the same way separately.

You now have a clean cape with perhaps a little fat still attached to the skin. Personally I leave this at this stage, since you can damage the cape by scraping with a knife. Some people advise that the cape should now be pinned out on a board to dry, but this is a slow business and I have not found it to be necessary. I pull the cape to the correct shape – cleaning can elongate the skin – and lay it skin side downwards on six or seven sheets of newspaper and leave it in a moderate or cool rather than a warm room. After twenty-four or forty-eight hours you will find that the newspaper has absorbed a lot of the moisture from the skin, but the cape is still quite soft. I lay the cape hackles down on a board and scraping with the edge of a sharp knife remove any borax, fat or membrane on the skin. Beat the cape against a wall to remove any borax from between the feathers, pull to shape and replace skin side down on fresh newspapers. Leave until quite dry, which will be in about a fortnight. If dried in too high a temperature, the cape tends to become hard and to turn up at the sides. I have never found that a cape treated in this way is attacked by moths or deteriorates.

If you do soil the hackles with blood, do not worry. Sponge clean with cold water or if necessary place the whole cape in a bowl of clean *cold* water; wash well; rinse under the cold tap and shake off the surplus water. Never use hot water or soap, which will remove the natural oil. Spread the bedraggled mass to shape and place *between* sheets of newspaper in a warm place. Change the newspapers daily until the feathers are dry when a firm shaking and gentle beating against a wall will miraculously transform them to their original state. The skin can then be dried in the normal manner.

Some people consider that pyrethrum is preferable to borax, leaving the skin in a softer condition. If you are preparing a few capes, I think that this is probably correct. Certainly the skins treated with pyrethrum that I have seen are very soft, but if you are preparing a large number of capes, you use quite a large quantity of drying powder and pyrethrum is much more expensive than borax. I now scrape all surplus borax off the skin after twenty-four/forty-eight hours drying, so the skin dries virtually without treatment. When dry, the skin is not too hard and I have found that within a year it becomes soft. Without borax or pyrethrum, I am sure that any powder with absorbent properties would act reasonably well. Household flour might be perfectly satisfactory.

Skinning a neck is one of these things which no two people will do the same way. I have given my method, but anyone doing it will quickly evolve their own. After quite a lot of practice, I can now skin and prepare a neck ready for drying in about five minutes.

The imported capes that you purchase commercially are always very much cleaner and freer from membrane than a home processed cape. The skin is almost paper thin and in some cases the skin on the side of the head has been cut round below, not above, the eye, the eye holes showing on the cape. For a long time, I could not understand how this was achieved. After skinning I tried careful scraping with a knife, but you cannot do it this way and if you continue too far you eventually loosen the hackles. I now consider that the way in which it may be done is to boil the head and neck. This allows the skin and the skin alone to be easily peeled away from the flesh, but boiling ruins the quality of the hackles since it removes the natural grease. There may be another secret method of achieving this result, but I feel sure that boiling is the answer. It makes skinning and preparing easy, if you can stand the smell, and it would also account for the unnatural soft feeling of the hackles, which is typical of a cape imported from the East.

Index

Index